T0192840

Fire Protection Engineering Applications for Large Transportation Systems in China

Fang Li • Huahui Li

Fire Protection Engineering Applications for Large Transportation Systems in China

Springer

Fang Li
Research Scientist
Worcester Polytechnic Institute
Worcester, MA, USA

Huahui Li
IKEA
Xuhui District, Shanghai, China

ISBN 978-3-030-58371-2   ISBN 978-3-030-58369-9   (eBook)
https://doi.org/10.1007/978-3-030-58369-9

This Springer imprint is published by the registered company Springer Nature Switzerland AG
The registered company address is: Gewerbestrasse 11, 6330 Cham, Switzerland

Contents

Chapter 1
Introduction

1.1 Definition of Large Transportation Hub

A large transportation hub is a significant constituent of a national or regional transportation system in cities such as airport and railway and supporting urban transport facilities such as rail transit station, public transport hub station, social parking garage, taxi business station, etc., which plays an important role and position in the urban passenger transport hub system.

The large transportation hub is the junction negotiating the routes of various transport modes. The installed equipment and mobile equipment together form a whole entity which serves as a hub to undertake the direct and transferring traffic and connect other cities. According to different transport modes, transportation hubs can be divided into rail hubs, bus hubs, airline hubs, and waterway hubs [1] (Fig. 1.1).

Social development has quickened the step of urbanization and led to urban population's sharp rise. With traditional small transportation hubs failing to meet the travel needs, people began to travel by plane and railways for convenience and budget reasons, and since then, some airports and stations developed into large contemporary transportation hubs. Tables 1.1 and 1.2 list some busiest airport terminals and railway stations in the world:

As the biggest financial center in China, Shanghai, being famous for Shanghai Disney-themed park, is rated as the world-level first-tier city by world authority GaWC. As one of the most populated cities in China, Shanghai has formed a large comprehensive transportation network composed of five transportation modes of railway, waterway, roadway, aviation, and rail. The Shanghai Hongqiao Comprehensive Traffic Hub, which was designed by East China Architectural Design Institute, is the first to integrate high-speed rail and airport worldwide. In 2018, Shanghai Hongqiao International Airport was titled as "Platinum Airport" by IATA and certificated as "five-star airport" in the Skytrax "2019 World Airport Competition" Award Ceremony held in London in 2019 (Fig. 1.2).

© Springer Nature Switzerland AG 2021
F. Li, H. Li, *Fire Protection Engineering Applications for Large Transportation Systems in China*, https://doi.org/10.1007/978-3-030-58369-9_1

California High-Speed Railway Station

Hartsfield-Jackson Atlanta Airport Terminal

Port Authority Bus Terminal

Port of Los Angeles

Fig. 1.1 Examples of transportation hubs

Table 1.1 Ten busiest airports in the world by passenger traffic in 2019 [1]

Airport	Location	Country	Number of passengers handled
Hartsfield-Jackson Atlanta International Airport	Atlanta, Georgia	United States	107.3 million
Beijing Capital International Airport	Chaoyang-Shunyi, Beijing	China	100.9 million
Dubai International Airport	Garhoud, Dubai	United Arab Emirates	89.1 million
Los Angeles International Airport	Los Angeles, California	United States	87.5 million
Tokyo International Airport	Tokyo	Japan	87.1 million
O'Hare International Airport	Chicago, Illinois	United States	83.3 million
Heathrow Airport	Hillingdon, London	United Kingdom	80.1 million
Hong Kong International Airport	Hong Kong	China	74.5 million
Shanghai Pudong International Airport	Pudong, Shanghai	China	74 million
Charles de Gaulle Airport	Île-de-France	France	72.2 million

Table 1.2 Ten busiest railway station in the world

	Location	Country	Number of passengers handled
Nanjing South Railway Station	Nanjing	China	32 million
Gare de Paris-Est	Paris	France	34 million
Shanghai Hongqiao Railway Station	Shanghai	China	52.7 million
Wuhan Railway Station	Wuhan	China	73.6 million
Munich Central Station	Munich	Germany	127.75 million
Zhengzhou East Railway Station	Zhengzhou	China	147 million
Guangzhou New Railway Station	Guangzhou	China	170 million
Grand Central Terminal	New York	United States	182.5 million
Tokyo Station	Tokyo	Japan	383 million

Fig. 1.2 Shanghai Hongqiao comprehensive traffic hub

With the gradual perfection of the passenger transportation and cargo transportation as well as various auxiliary facilities around Hongqiao comprehensive traffic hub, the huge passenger traffic attracted by the station not only bring along the tourist and business market along the city but also give more chance to various elements to flow; as a result, many city resources witnessed a further optimization and integration around the Hongqiao traffic hub, forming a city subcenter with the traffic hub at its core (Fig. 1.3).

The rise of large transportation hubs has posed unprecedented difficulties and challenges to fire protection design. The book will collect data about several large transportation hubs, make investigation into them, explore the spatial organization types, and make a summarization. Departing from performance-based fire

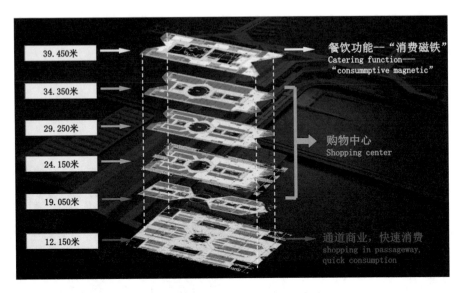

Fig. 1.3 The principle of commercial layout in Hongqiao traffic hub

protection, using a digital simulation of fire smoke diffusion and people evacuation, it will analyze the data result and discuss evacuation scenarios and eventually explore further to find better approaches for safety design of large transportation hubs.

1.2 Large Transportation Hubs Accelerate Urban Development

Transport infrastructure is the precondition for a well-functional city as well as its economic and social development. Good transport infrastructure promotes a city's economic development by creating more opportunities to connect surrounding areas and other cities in the world and helps maintain its long-standing position as a transportation hub. Internationally famous transportation hubs include Chicago, Frankfurt, Dubai, and Hong Kong.

1.2.1   The Development of Internationally Famous Transportation Hubs

1.2.1.1   Chicago

Located in the heartland of the North American continent, Chicago is an important railway and airline hub, apart from being a financial, cultural, manufacturing, futures, and commodity trading center.

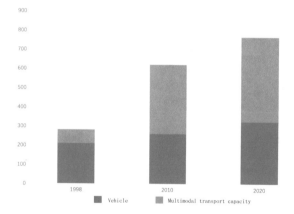

With a sprawling transport networks, entitled America's Arteries, Chicago is a hub for waterway, railway, and air transportation. Via Saint Lawrence River and Mississippi River, it has direct accesses to Europe and Mexico Gulf. As the most important node on American railway network, it negotiates the central and northern lines, with 37 main lines and 35,000 carriages coming and going daily. Its total mileage of 12,400 km and annual freight amount of 512 million tons both top other major cities in the world. Chicago has two international airports, O'Hare International Airport and Midway International Airport. As one of the busiest airports in the world, O'Hare International Airport is in the front rank in the United States as well as in the world in terms of plane traffic, passenger numbers, and cargo tonnages and receives more than 69 million passengers annually.

It is a world-class financial center; home to many superlative financial institutions; the biggest exchange for damageable goods; the biggest, earliest future and option trading exchange; and the futures exchange holding a record high turnover. Some large banks or financial setups also headquarter or set up branches here. Over 300 US banks, over 40 branches of foreign banks, and over 16 insurance companies have settled here, among which 33 are listed in Fortune 500 and 47 are Forbes 500 firms.

1.2.1.2 Frankfurt

Frankfurt is located in central Europe and the heartland of Germany. Defined by a favorable location, Frankfurt has developed into an important transportation hub and business, financial, and international convention center of Germany and Europe.

Frankfurt is at the crossroads of the European continent, serving as a transportation hub for Rhine-Main and Europe. Frankfurt is one of the most important railway nodes in Europe and the only passage leading from western European countries to Munich. In Europe, Frankfurt Airport has the largest freight amount, while its passenger capacity is only second to London Heathrow International Airport, which undoubtedly makes it an internationally important air transportation hub. The flights in Frankfurt Airport fly to and from five continents, and all international airlines bound for European countries transfer in Frankfurt, although the departure cities also have direct flights. Well-equipped railway and airline infrastructure, a transport network covering Europe and connecting the world, these laid foundation for Frankfurt's role as an international transportation hub.

In the center of Germany and at the junction of European transport lines, the nice geographic location attracts European Central Bank and Deutsche Bundesbank to Frankfurt and makes it Europe's financial center and one of the four financial centers in the world. Frankfurt Stock Exchange is one of the world's largest exchanges. Numerous financial institutions and the booming financial market solidify its position of a global financial center.

Frankfurt is the busiest convention center in Europe. Every year, over 50,000 meetings were held here attracting more than 2.6 million senior executives, among which 15 are prominent, including the International Consumer Goods Fair, the International Sanitary Ware Heating and Air-Conditioning Expo, the Heimtextil, the International Automobile Exhibition, the Cooking Appliances Exhibition, etc. It ranks the first in the world in terms of the numbers, attendances, and grades of the meetings. This is also reflected in the comparison of the core indicator data of Frankfurt and the international convention cities of Munich and Shanghai, as shown below (Table 1.3).

Table 1.3 Comparison of core indicators for the exhibition industry in Munich, Frankfurt, and Shanghai

Indicators		Explanation	Munich	Frankfurt	Shanghai
Exhibition hall	Area of the hall	Total area of main hall (unit:10 thousand m²)	44	41	35.24
Product to be display	Area for exhibition	Scale factor of offshore rental area	31.64	51.26	27.81
	Number of participating companies	Scale factor of offshore participating companies	49.40	59.90	24.80
	Number of audiences	Scale factor of foreign audiences	32.34	35	6.4
Involving companies	Number of professional exhibiting company	Number (unit: company)	178	178	74
	Scale of professional exhibition company	Population (unit: occupant)	3121	3121	3700
	Number of practitioner	Total volume of practitioner (unit:10 thousand occupants)	2	2	3
	Number and scale of brand exhibition	Number of exhibition passing UFI certification (unit: times)	19	15	17

1.2.1.3   Dubai

As the largest city of the United Arab Emirates and the economic and financial center of the Middle East or even of the world, Dubai bridges the eastern and western capital markets.

Dubai Airport is the largest in the Middle East, and in 2018, its passenger throughput reached 89.14 million and ranked the third in the world while with a second largest international passenger throughput, which established its position as an air transportation transfer hub for Europe, Africa, and Asia. As one of the most important air transportation hubs in the world, flights departing from Dubai Airport fly to 239 cities in North America, Europe, South America, East Asia, Southeast Asia, South Asia, Oceania, and Africa.

With the support of the airport, Dubai's influence as a financial center has expanded from the region to the world. Adopting a liberal and steady economic policy, Dubai enjoys a great reputation among all countries and in the international industrial and commercial circles, and it is directing the domestic and foreign capital to the commercial, industrial, and service industries (Table 1.4).

Dubai is rich in tourism resources: a 64 km long beach, well-equipped resort hotels plus sailing, water skiing, surfing, diving, fishing, golfing, desert safaris, bird watching, and other recreational activities. The tourism boom of Dubai is driven by the growth of its aviation industry. Oil only makes up 6% of its GDP, most of which comes from tourism.

Table 1.4   Ranking of global airport throughput in 2018

Rank 2018	Rank 2017	Airport city/Country/Code	Passengers (Enplaning and deplaning)	Percent change
1	1	Atlanta GA, US (ATL)	107,394,029	3.3
2	2	Beijing, CN (PEK)	100,983,290	5.4
3	3	Dubai, AE (DXB)	89,149,387	1.0
4	5	Los Angeles CA, US (LAX)	87,534,384	3.5
5	4	Tokyo, JP (HND)	87,131,973	2.0
6	6	Chicago IL, US (ORD)	83,339,186	4.4
7	7	London, GB (LHR)	80,126,320	2.7
8	8	Hong Kong, CN (HKG)	74,517,402	2.6
9	9	Shanghai, CN (PVG)	74,006,331	5.7
10	10	Paris, FR (CDG)	72,229,723	4.0
11	11	Amsterdam, NL (AMS)	71,053,147	3.7
12	16	New Delhi, IN (DEL)	69,900,938	10.2
13	13	Guangzhou, CN (CAN)	69,769,497	6.0
14	14	Frankfurt, DE (FRA)	69,510,269	7.8
15	12	Dallas/Fort Worth TX, US (DFW)	69,112,607	3.0
16	19	Incheon, KR (ICN)	68,350,784	10.0
17	15	Istanbul, TR (IST)	68,192,683	6.4
18	17	Jakarta, ID (CGK)	66,908,159	6.2
19	18	Singapore, SG (SIN)	65,628,000	5.5
20	20	Denver CO, US (DEN)	64,494,613	5.1
TOP 20 FOR 2018			1,539,332,722	4.7

1.2.1.4   Hong Kong

Table 1.5 Ranking of global port throughput in 2018

Rank	Port	2018 mTEU	Growth
1	Shanghai	42.01	4.40%
2	Singapore	36.6	8.70%
3	Ningbo	26.35	7.10%
4	Shenzhen	25.74	2.10%
5	Guangzhou	21.87	7.40%
6	Busan	21.66	5.70%
7	Hong Kong	19.6	−5.7%
8	Qingdao	19.32	5.50%
9	Los Angeles/Long Beach	17.55	3.90%
10	Tianjin	16.01	6.20%

Hong Kong backed by the Mainland China, facing Southeast Asia, at the heart of the Asia Pacific Region and the fast-growing Asia Pacific Rim, negotiates the south and north passages of the Asia Pacific Region and serves as an important transportation hub for China and Southeast Asian countries. It is also an important financial center in Asia.

As one of the most important air transportation hubs in Asia and the world, Hong Kong has one airport that is of the busiest with the fifth largest international passenger capacity and the largest international freight amount. Port of Hong Kong leads the world in terms of cabin tonnage, cargo throughput, and passenger throughput. For a very long time, its container throughput ranks No.1 in the world. At present, it has more than 20 sea routes, more than 80 international liners, and around 500 times of weekly container liner services. Departing from Port of Hong Kong, the liners are bound to about 1000 ports in over 120 countries and regions. It is the most important water transportation hub in Asia and a key port along the global supply chain. With an excellent location, a well-equipped airport, and a port connecting the air and water transportation routes, Hong Kong is an ideal transportation and transfer hub to connect Mainland China with foreign freights and passengers, apart from being an internationally renowned transportation hub (Table 1.5).

Second only to London and New York, Hong Kong is the third largest financial center in the world. It has the most concentrated foreign banks and highly internationalized banking system and businesses. The efficient stock market in Hong Kong is of high liquidity and with a sound risk management system. As one of the ten largest stock markets in the world, the market value in Hong Kong closely follows Tokyo in Asia. Hong Kong's foreign exchange market plays a dominant role in the Asia Pacific region, and its average daily trading volume ranks the fifth in the world; the average trading volume of foreign exchange derivatives also ranks the fifth internationally, while the daily trading volume of interest rate derivatives is in the eighth place in the world.

The previous and present development of the abovementioned nodal cities reveals that a city's role as a transportation hub and its economic, financial, and cultural development is inseparable. Transportation is the foundation, guide, and guarantee for economic and social development, and the economic, financial, and commercial development will accelerate the concentration of people and commodities, which will propel social and economic development. It is worth mentioning that air transportation hubs play a significant part in negotiating all means of transportation in key transportation nodes and removing the limit of spaces and further promoting the influence of logistics-based cities.

1.2.2 The Development of Main Logistics-Based Transportation Hubs in China

A city's economic development is closely related to its geographic location. Many cities serve as important regional transportation hubs, and the development of metropolises such as New York and London is supported by their geographic locations. Major regional transportation hubs can attract and disperse all kinds of essential productive factors in the peripheral regions and influence the development of the regional economic development. With the growth of the economy of transportation hubs comes the improvement of the traffic conditions, which in turn has a greater influence on the regional economic development. Since the reform and opening up, China's economy has been soaring, and many ports and inland transportation hubs have taken advantage of their geographic locations and achieved great success. Shanghai, Beijing, Guangzhou, and a series of well-developed transportation hubs have become the pillars for China's economic development, led the surrounding areas into fast growth, and formed some urban agglomerations along the Yangtze River, Beijing-Tianjin-Hebei economic zone, and other areas.

The following part is a summary of the development of some major transportation hubs in China.

1.2.2.1 Shanghai

Shanghai is the financial center of China as well as the second one to have a Disney-themed park in China, with the first one being Hong Kong. On March 16, 2010, Terminal 2 at Hongqiao Airport was opened, and Hongqiao Integrated Transportation Hub became the largest in China to successfully operate railway, air, and ground transportation. The entire hub spans a total area of 1.2 million square meters, including the terminals, high-speed railways, a maglev, and five subway lines linking the east and west Shanghai. The hub provides 64 connecting ways and 56 transfer ways for passengers, with a daily average passenger flow of 1.1 million. In 2015, the terminals are expected to receive 40 million passengers and 300,000 planes annually, and 16 platforms and 30 tracks will be used by high-speed and inter-urban trains. In 2020, it is expected to serve 52.72 million passengers annually, and in 2030, the number will grow to 78.38 million (Fig. 1.4).

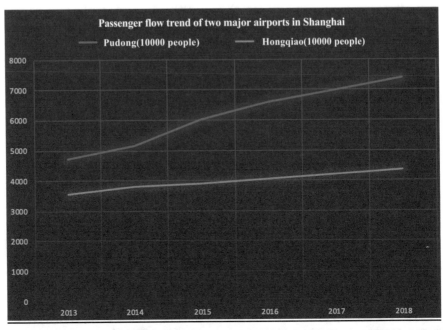

Fig. 1.4 Shanghai Hongqiao comprehensive traffic hub

Shanghai Pudong International Airport is located 30 km away from city center at Pudong New District. It has two terminals and three cargo areas, with 218 parking bays, 135 of which are used by passenger planes. In 2017, the annual passenger throughput of Pudong Airport reached 70.0043 million, annual cargo throughput

Fig. 1.5 Shanghai Pudong comprehensive airport

Fig. 1.6 Shanghai Pudong satellite hall

3.8356 million tons, and annual flights 496,879. As of the end of 2017, 110 airline companies had regular flights to the two airports of Shanghai, and Pudong Airport had linked 297 aviation points in 47 countries and regions (Fig. 1.5).

On December 30, 2015, the satellite hall started construction, which was the main building of the third phase of the extension project. Covering a total floor area of 622,000 sqm, it is the biggest single-building satellite hall in the world and will be put into use by the end of 2019 (Fig. 1.6).

1.2.2.2 Beijing

While as the capital, political center, cultural center, international exchange center, and the technological innovation center of China, Beijing is a world famous ancient capital as well as a modernized international city and renowned for some scenic spots like the Palace Museum and the Great Wall. As one of the centers of China railway network, the Beijing Nan Railway Station, which is the second large one in Asia, occupies an area of 320 thousand m² and has 24 rail lines and 13 platforms.

Terminal 3 is located on the east of Beijing Capital International Airport. The main building and its auxiliary projects are distributed in between the east runway and the new runway. It is the second largest single-building terminal in the world. It consists of the main building and domestic and international concourses and equipped with automatic processing and high-speed baggage system, easy passenger transit system, and information system, with a total floor area of 986,000 sqm.

Beijing Daxing International Airport has been completed in 2019. Situated between Daxing District of Beijing and Guangyang District of Langfang, Hebei Province, it is a mega international airport and an integrated transportation hub. In 2040, its passenger and plane throughputs are expected to reach 100 million and 800,000, respectively, with 7 runways and 1 terminal with an area of 1.4 million sqm. And in 2050, passenger and the plane throughputs are expected to reach 130 million and 1.03 million, respectively, with 9 runways.

The main project of Beijing Daxing International Airport is mostly in Beijing, which is the third passenger airport following Beijing Capital International Airport and Beijing Nanyuan Airport (to be relocated). The present phase includes 4 runways, a dual-use runway (new Nanyuan Airport for the air force), a 700,000 sqm terminal, and 92 parking bays for passenger planes. It is expected that, when completed by the end of 2019 and the passenger throughput reaches 45 million, the first satellite hall will be built; by then, the area of the terminal will reach 820,000 sqm, and the number of parking bays will increase to 137 with a design capacity of 72 million passengers.

1.2.2.3 Hangzhou

Hangzhou as the capital of Zhejiang Province is located in the coastal area of southeastern China. It was the host city of the 11th G20 summit and the place where the word famous e-commerce enterprise Alibaba is located. As an important traffic hub of China, Hangzhou Dong Railway Station has the volume equivalent to Shanghai Hongqiao Traffic Hub. Hangzhou Dong Railway Station covers 40 hectares (600 mu), with a total floor space of 1 million sqm (350,000 aboveground and 650,000 underground); it is an essential part for Hangzhou's railway

system and one of the largest junction stations in Asia. Hangzhou East Railway Station has two station squares, one on the east and the other west, which take up an area of 488 mu with a gross floor area of 813,000 sqm. The east station square takes up an area of 185 mu with a 160,000 sqm aboveground floor area and a 210,000 underground floor area, and the west station square takes up an area of 300 mu with a 110,000 sqm aboveground floor area and a 330,000 underground floor area.

Phase 3 project of Hangzhou Xiaoshan International Airport covers 690,000 sqm, almost twice the size of Terminal 1, Terminal 2, and Terminal 3 combined. As the new portal to-be for Xiaoshan International Airport, together with the existing terminals, it will become a transportation center and make up an integrated transportation hub combining "air, subway, and railway transportation modes." Upon completion, it is estimated that, in 2030, it will be able to meet the travel needs of 90 million passengers. With the addition of two mid-range runways on both sides of the existing ones, the total runway area is set to reach 200,000 sqm. In the future, Xiaoshan International Airport will be worthy of the name "a hub for Zhejiang" and a postcard of the province's image.

The construction and development of transportation facilities in abovementioned cities are a microcosm of major hub cities in China. Guangzhou, Tianjin, Chengdu, Wuhan, Xi'an, and many other cities all have large-scale high-speed railway stations and are even planning a second airport. The development of Chinese cities is largely dependent on constructing transportation hubs. In addition, second- and third-tier cities in China are beginning to build large transportation hubs. The fast growth of high-speed trains is contributing to the development, prosperity, and steadiness of a region's economy, the optimal allocation of resources, and the realization of urbanization. China has built 29,000 km of high-speed railway – accounting for two thirds of the world's total, more than the rest of the world combined. Table 1.6 shows the ten busiest high-speed railway stations in China:

Table 1.6 Ten busiest high-speed railway stations in China

Ranking	Station	Scale	Floor area (10,000 sqm)	Investment (100 million yuan)
1	Xi'an North Railway Station	18 platforms and 34 tracks	42.5	61
2	Zhengzhou East Railway Station	16 platforms and 34 tracks	41.2	94.7
3	Shanghai Hongqiao Railway Station	16 platforms and 30 tracks	24	150
4	Kunming South Railway Station	16 platforms and 30 tracks	33.4	31.8
5	Guiyang North Railway Station	15 platforms and 32 tracks	25.5	66.7
6	Chongqing West Railway Station	15 platforms and 31 tracks	30	30.8
7	Hangzhou East Railway Station	15 platforms and 30 tracks	34	98
8	Guangzhou South Railway Station	15 platforms and 28 tracks	33.6	130
9	Nanjing South Railway Station	15 platforms and 28 tracks	45.8	140
10	Chongqing North Railway Station	14 platforms and 29 tracks	26.6	Renovated

1.3   The Spatial Patterns of Transportation Hubs

1.3.1   The Spatial Patterns of Large Modern Railway Hubs

The architectural forms and spatial patterns of railway hubs generally base on how passengers enter and exit a station, and there are three patterns to follow: "enter above and exit below," "enter and exit in parallel," and "enter below and exit below." The two patterns, "enter above and exit below" and "enter below and exit below," use elevated ramps to enter and exit at ground level or enter and exit both at ground level. Elevated ramps solve the problem of high-speed trains entering or leaving at the same direction and provide the passengers with easiest accesses to trains, which is common in large transportation hubs. The "enter and exit in parallel" pattern is more common in medium- and small-scale railway stations. Table 1.7 shows enter and exit patterns of passengers in some high-speed railway stations:

Since the subject of the book is large transportation hubs, the introduction of enter and exit patterns will mainly focus on large railway stations:

① "Enter above and exit below" pattern:

Taking Shanghai Hongqiao transportation hub, for example, it adopts an "enter above and exit below" pattern; all enter facilities are set above the station site, and passenger will take the elevated ramp to enter and exit at the ground or underground levels, which realizes the stratification of passenger flows.

(1) The flow of entering passengers

Table 1.7 Enter and exit patterns of passengers in some high-speed railway stations

Station	City	City scale	Station scale	Station type	Enter and exit patterns of passengers
Shanghai Hongqiao Railway Station	Shanghai	Mega	Mega	Crossed	Enter above and exit below Enter below and exit below
Nanjing South Railway Station	Nanjing	Mega	Mega	Crossed	Enter above and exit below
Guangzhou New Railway Station	Guangzhou	Mega	Mega	Crossed	Enter above and exit below Enter below and exit below
Chengdu New Railway Station	Chengdu	Mega	Mega	Crossed	Enter above and exit below
Beijing South Railway Station	Beijing	Mega	Mega	Crossed	Enter above and exit below Enter below and exit below
Wuhan New Railway Station	Wuhan	Mega	Mega	Crossed	Enter above and exit below
Qingdao Railway Station	Qingdao	Large	Medium	End to end	Enter and exit in parallel
Shenyang Railway Station	Shenyang	Large	Medium	Crossed	Enter and exit in parallel
Foshan Railway Station	Foshan	Large	Medium	Crossed	Enter below and exit below

Passengers enter entrance hall of the high-speed railway station and ticket hall of the airport via the east and west overpasses and arrive at waiting rooms or departures after passing through the security.

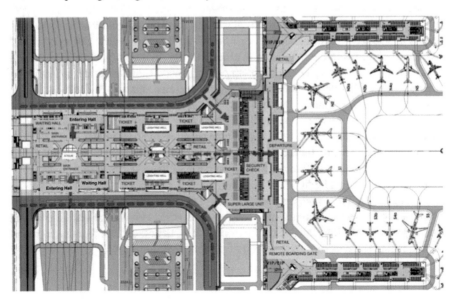

(2) The flow of leaving passengers

Air travelers can leave the station or transfer to other transportation means by way of "no luggage" passage, and railway passengers can leave the platform via exit hall.

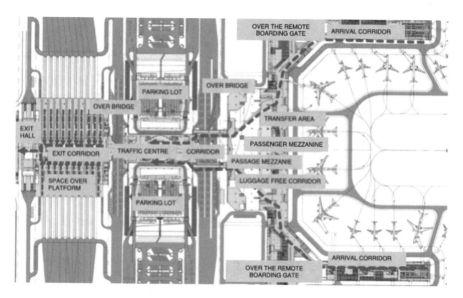

(3) The flow of transferring passengers

Passengers can use the vertical channels and change to other transportation means (Fig. 1.7).

① "Enter below and exit below" pattern

Taking Foshan West Railway Station, for example, it adopts an "enter below and exit below" pattern. All enter and exit facilities are set below the station site shown as Fig. 1.8, and passenger will enter at the ground and exit at the interlayer. And the stratification of passenger flows is realized under the bridge. This entering pattern helps with convergence and divergence of travelers of different transportation means [2].

(1) The flow of entering passengers

The passengers on foot or by bus coming from the north and south station squares can enter directly by the north and south entrances after security check.

Passengers coming by taxi get off at the east and west drop-off zones and walk along the entrance platform and entrance hall to reach the waiting rooms.

Passengers coming by private cars get off at the parking lot on the northwest and walk along the entrance platform and entrance hall to reach the waiting rooms.

Passengers coming by buses get off on the southeast bus station and enter the station by east or south entrance.

Passengers coming by subway can use elevators and escalators to go upward to the enter platform.

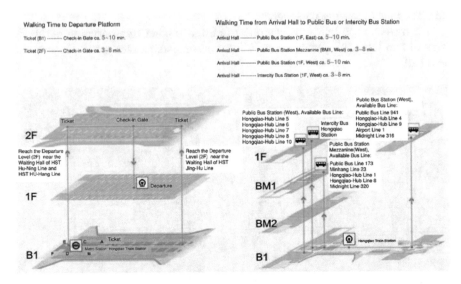

Fig. 1.7 Sketch map of transferring plan at Shanghai Hongqiao transportation hub

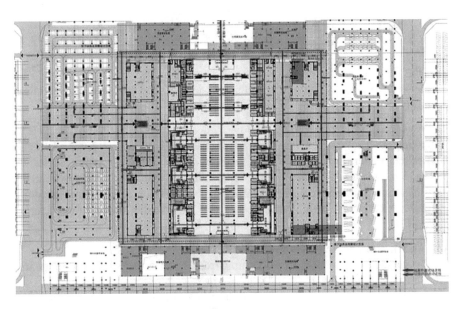

Fig. 1.8 Sketch map of entering passenger flow

(2) The flow of entering passengers

Long-distance passengers enter the station through three check-in gates on each side of the waiting hall, enter the entrance corridor, and then ride the corresponding escalator or elevator to the platform.

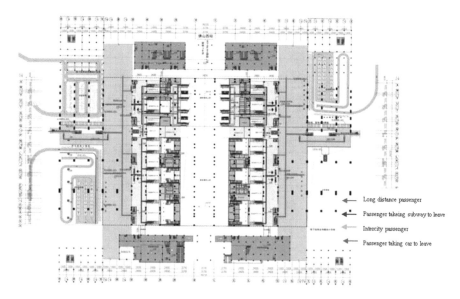

Fig. 1.9 Sketch map of leaving passenger flow

Intercity passengers enter the station through two check-in gates on each side of the waiting hall, ride escalators or elevators to the entrance corridor of the exit floor (i.e., the interlayer on the first floor), and then ride the corresponding escalator or elevator to the platform.

After ticket check, travelers with special needs may use the barrier-free elevator to go directly to the corresponding platform.

(3) The flow of leaving passengers (Shown as Fig. 1.9)

Long-distance passengers can go down to the exit corridor on the exit floor (the interlayer on the first floor) by escalator and leave through the passenger exit platform after ticket check. Travelers with special needs may use the barrier-free elevator to go directly to the entrance corridor on the entrance floor and leave through the waiting room after ticket check.

Intercity passengers can leave the platform by escalators and go down to exit hall (the interlayer on the first floor) on the exit floor and leave after ticket check. Travelers with special needs may use the barrier-free elevator to go directly to the entrance corridor on the entrance floor and leave through the transfer passage for intercity travelers.

Passengers can walk along the exit platforms to the south or north square and leave the station. Passengers on the east outbound platform take the nearest bus and taxi to leave the station, while those on the west outbound platform take the nearest taxi and self-contained car to leave the station. The passengers taking car use the overbridge crossing over East Zhangchang Road and West Zhangchang Road in and reach the under bridge throat area, where locates a large-scale parking lot for private

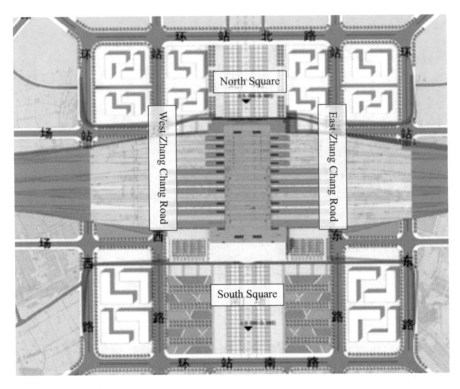

Fig. 1.10 Overall layout of Foshan railway station

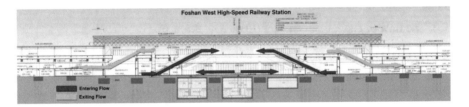

Fig. 1.11 Sketch map of the entering and exiting flow of long-distance passengers

cars. Passenger taking subway use elevators and escalators to get to the entrance platform and leave the station by subway through the subway or underground entrance (Fig. 1.10).

(4) The flow of transferring passengers (shown in Fig. 1.11)

Long-distance passengers and intercity passengers can transfer without leaving the station. They can walk along the transfer passage of the exit hall; pass through the exit hall, the transfer passage, and the entrance corridor for intercity passengers; and go to the corresponding platform after ticket check.

Intercity passengers can transfer by way of exit hall and the transfer passage and reach the platform via entrance corridor, or they can pass the transfer passage of exit

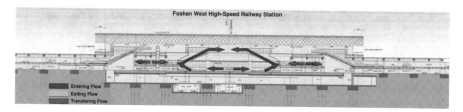

Fig. 1.12   Sketch map of the entering and exiting flow of inter-urban passengers

Fig. 1.13   Sectional drawing of Shenzhen Futian Railway Station

hall, reach the long-distance section, and go to the corresponding platform after ticket check.

Recently, through the south and north activity platforms and the underground space, the convergence of passengers coming from east and west platforms can be achieved. In the future, after the completion of the south and north auxiliary buildings, the convergence of passengers coming from east and west platforms can be achieved through the exit platform and the auxiliary buildings (Fig. 1.12).

③ The entering and exiting flow of passengers at large underground railway stations

Taking Shenzhen Futian Station for an example, it is a three-tier building and covers an area of 147,000 sqm. B1 is a transfer hall with 16 entrances and exits; B2 is where station halls and waiting rooms locate, with 4 check-in gates at the entrance and more than 1200 seats capable of holding 3000 passengers; and B3 is the platform floor, with 8 tracks and 4 platforms. Passengers can seamlessly transfer to subways, buses, and taxis (Fig. 1.13).

1.3.2 The Spatial Patterns of Large Modern Airport Terminals

Fundamentally, to design an airport terminal is to arrange its layout in accordance with its scale, planning conditions, runway and taxiway plans, and operation models, which decide the overall design of the terminal. This is the most important design features of every airport terminal, which determines the operating mode of the terminal and the arrangement of space, buildings, and passages.

For terminal buildings, the purpose is to think of a reasonable design and form appropriate spatial patterns. In large airports, at present stage, the focus is to reduce the walking distance of the passengers and give clear directional information; the space of the terminal areas has to hold enough parking aprons and ensure efficient gliding.

The land planning, the terminal square, and the overpasses combine to form a good transportation system and guarantee the connection of terminals and land traffic. This shows that the structural design of a terminal greatly influences its spatial and process organization. Therefore, the study of the space and process of a terminal begins with a profound understanding of its structural pattern.

By conducting researches on the recent design of terminals, the model can be summarized into three main types: linear terminals, pier-shaped terminals, and satellite terminals.

1.3.2.1 Linear Terminals

Linear terminals consist of the main building and the departure hall, with planes waiting on one side and passenger embarking on and disembarking from the aircrafts through a boarding bridge. This compact layout shortens the walking distance of the passengers with a relatively clear process. And the long curbside platforms make it easier for expansion. The shortcoming is limited passenger capacity, and expansion is an uneconomical choice and won't accommodate the large flow of the terminals. It has two development modes: one mode is the decentralization of handling areas and waiting areas, which forms several waiting units and a large terminal with the road transportation, and Hamburg Airport in Germany and Dallas/Fort Worth International Airport in the United States are examples of this development mode; the other mode is the concentration of handling areas and the decentralization of waiting areas, which enhances the check-in and baggage handling functions, extends the departure and arrival halls, and makes room for more planes (Figs. 1.14 and 1.15).

1.3.2.2 Pier-Shaped Terminals

A pier-shaped terminal consists of the main building, responsible for check-ins and luggage handling and one or several corridors serving as waiting areas. And it has longer station sites and can accommodate more planes. Long walking distance, divergence of overcrowded passenger flows, and inconvenience for plane to operate,

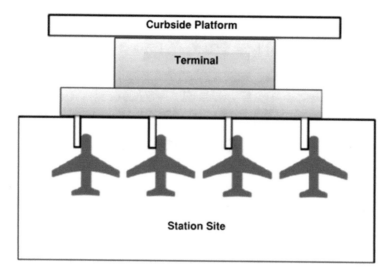

Fig. 1.14 Linear terminals

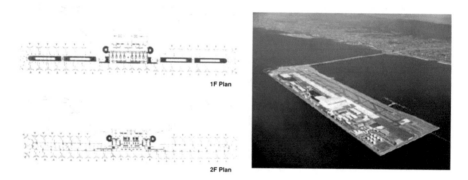

Fig. 1.15 Conformation of various kinds of terminals

these are the problems of pier-shaped terminals. But they are very common both in China and abroad, such as Guangzhou Baiyun International Airport, Xi'an Xianyang International Airport, Incheon International Airport in South Korea, and Bangkok International Airport in Thailand, for the reasons of easy extension and economic advantages (Figs. 1.16, 1.17, and 1.18).

1.3.2.3   Satellite Terminals

Satellite terminals separate waiting rooms from the main building and are connected by ground and underground transportation. Passengers check in at the main building, while the operation of planes is relatively flexible and not restricted by the main

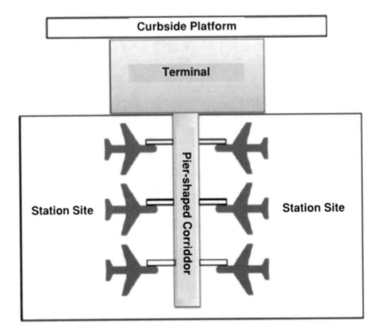

Fig. 1.16 Pier-shaped terminals

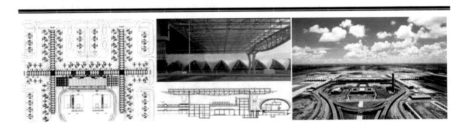

Fig. 1.17 Bangkok International Airport

Fig. 1.18 Guangzhou Baiyun International Airport

building. The departure halls are easy to extend, but the travel distance within the terminals is too long, and the station sites take up a very large floor area. This pattern is suitable for large and super large terminals, such as the terminal of Kuala Lumpur International Airport and T5 of London Heathrow Airport (Figs. 1.19 and 1.20).

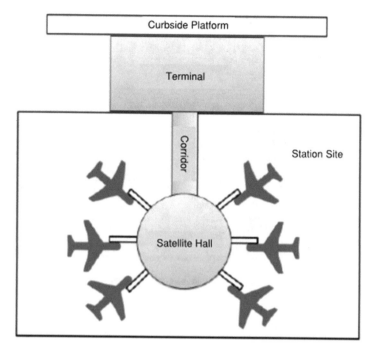

Fig. 1.19 Satellite terminals

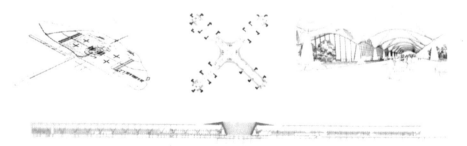

Fig. 1.20 Kuala Lumpur International Airport

These are the three basic conformation patterns of terminal.

In practice, they are combined to achieve the utmost efficiency. The new Kunming Airport and the T3 of Shenzhen Bao'an International Airport adopted a linear main building, while the wings and the back are in pier shapes. An ideal conformation combines the basic types and makes site-specific innovations. And it's expected that more new conformations of terminals will come into being in the future.

Fig. 1.21 Sectional view of Chongqing Shapingba Railway Station

1.3.3   The New "Multistory and Underground" Constructional Forms

The fast development of high-speed railway has presented opportunities for the upgrading of old railway stations in the central urban areas. With the urbanization and economic growth, transportation hubs of simple functions are transforming into comprehensive and integrated ones. The "multistory and underground" development ideas help save urban land resources and maximize people's activity spaces.

Above Chongqing Shapingba Railway Station rise the 180-meter high twin towers for property management, and the eight underground floors are for high-speed railways, subways, buses, and taxis (Fig. 1.21).

1.3.4   Underground High-Speed Railway Station

Underground railway stations are a new type of railway stations and transportation hubs.

The first-generation underground railway stations place the tracks and transfer function in the underground space, keep station building on the ground, and provide many advantages, such as blocking noise of the train, hiding the track underground, developing the track areas, and avoiding the separation of urban space and transportation by on the ground tracks, multidirectional tracks, and the change from waiting mode to transferring mode, which have a huge influence on the arrangement of key urban spaces [3] (Fig. 1.22).

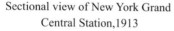

Sectional view of New York Grand Central Station,1913	Three-dimensional transportation of Berlin Central Station

Fig. 1.22   First-generation underground railway station

Sectional view of Yujiabao Railway Station	Sectional view of Shenzhen Futian Railway Station

Fig. 1.23   Second-generation underground high-speed railway station

The second-generation underground railway stations make efficient use of central urban space with the cancellation of station squares and the decrease of station waiting rooms. They also optimize the transportation organization of central urban space, such as improving the travel efficiency of mid- and long-distance travelers, promoting the upgrading of public environment and the establishment of underground pedestrian walk networks, and enhancing the advantages brought about by underground railway stations to the cities' core areas. With the acceleration of urbanization, the efficient, compact, and coordinated new form of underground railway stations will have a greater prospect. The purpose of this research is to fill the void in theories concerning the development of underground railway stations and urban centers (Fig. 1.23).

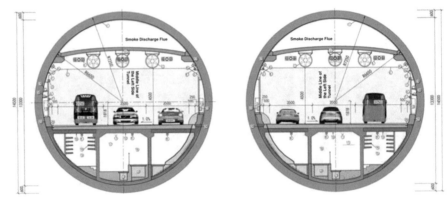

Fig. 1.24 Sectional view of shield tunnel

1.3.5 Ultra-Long Shield Tunnel

Transportation tunnels are an important component of urban transportation, which include the railway tunnels, highway tunnels, underwater tunnels, underground railways, and pedestrian tunnels. Shield tunnels are constructed with shield tunneling methods using shield machines. During the operation, operators have to take control of the excavation face and surrounding rock to avoid collapses caused by instability while excavating and removing slags. And they have to assemble duct pieces in the machine to make them into tunnel linings, grout behind lining, and construct a tunnel without disturbing surrounding rocks. The following table lists the largest urban transportation tunnels in China (Fig. 1.24 and Tables 1.8, 1.9, and 1.10):

1.4 Fire Hazards in Large Transportation Hubs

Large transportation hubs have become the landmarks and postcards of modern cities, with a high population density. The substantial increase of transportation capacity and speed leads to enlarged scales and expanded functionalities. The characteristics of fire hazards in large transportation hubs are reflected by the following aspects:

(1) Large fire load

A single floor of a large transportation hub can have an area of tens of thousands of sqm. Many waiting rooms at railway stations or airports can be as large as ten thousand sqm or more, and they are getting larger and larger. For example, Guangzhou New Railway Station covers an area of 760,000 sqm, with a total area of 458,000 sqm. With the integration of passenger services and freight service in modern transportation hubs, a large number of goods awaiting shipment and passenger luggage, mainly clothes, are present at the same time in the stations, which

Table 1.8 Ultra-long highway tunnels in China mainland

No.	Name	Length (m)	Location	Lanes	Ventilation type
1	Qinling Zhongnan Mountain highway tunnel	18,020	Shaanxi Province	2 × 2	3 shafts, segmented, vertical
2	Dapingli highway tunnel	12,290	Gusu Province	2 × 2	2 shafts, segmented, vertical
3	Baojia Mountain highway tunnel	11,500	Shaanxi Province	2 × 2	3 shafts, segmented, vertical
4	Baota Mountain highway tunnel	10,391	Shanxi Province	2 × 2	Inclined shafts, in and out, vertical
5	Niba Mountain highway tunnel	9985	Sichuan Province	2 × 2	Inclined shafts+ vertical, segmented
6	Mayazi highway tunnel	9000	Gansu Province	2 × 2	In and out through inclined shafts +jet fan, vertical
7	Longtan highway tunnel	8700	Hubei Province	2 × 2	In and out through holes+ jet fan, vertical
8	Mixiliang highway tunnel	7923	Shaanxi Province	2 × 2	In and out through one shaft on the left (right)
9	Kuocang Mountain highway tunnel	7930	Zhejiang Province	2 × 2	Vertical +semi-cross flow (smoke discharge)
10	Fangdou Mountain highway tunnel	7581	Chongqing	2 × 2	2 inclined shafts, in and out, vertical

Table 1.9 Major underwater highway tunnels in China

No.	Name	Length (m)	Location	Lanes	Ventilation type
1	Xiamen underwater tunnel (drill and blast)	5960	Fujian Province	3 × 2	In and out through shafts + jet fan, vertical
2	Shanghai Yangtze River tunnel (shield)	8955	Shanghai	3 × 2	Horizontal
3	Wuhan Yangtze River tunnel (shield)	3630	Hubei Province	2 × 2	Horizontal
4	Shangzhong road tunnel (shield)	2800	Shanghai	2 × 2	Horizontal (two-layer, two-way)
5	East Fuxing road tunnel (shield)	2785	Shanghai	3 × 2	Horizontal (two-layer, two-way)
6	Nanjing Xuanwu Lake tunnel (shield)	2660	Nanjing Province	3 × 2	
7	Dalian road tunnel (shield)	2566	Shanghai	2 × 2	Horizontal
8	Shanghai outer ring cross-river tunnel (immersed tube)	2882	Shanghai	4 × 2	Vertical
9	Pearl River tunnel (immersed tube)	1238	Guangdong Province	3 + 3	Vertical (for highways and railways)
10	Ningbo Changhong tunnel (immersed tube)	1053	Zhejiang Province	2 × 1	Vertical

Table 1.10 Long-span highway tunnel in China

No.	Name	Length (m)	Location	Lanes × tunnel holes
1	Baihezui tunnel	1240	Chongqing	4 × 2
2	Longtou Mountain tunnel	1020	Guangdong Province	4 × 2
3	Wanshi Mountain tunnel	1170	Fujian Province	Underground interchange with a widest point of 25.89 m
4	Dage Mountain tunnel	496	Guizhou Province	4 × 1
5	Jinzhou tunnel	521	Liaoning Province	4 × 1
6	Yabao tunnel	260	Guangdong Province	4 × 2
7	Jinji Mountain tunnel	200	Fujian Province	4 × 2 (arcade)
8	Luohan Mountain tunnel	300	Fujian Province	4 × 2 (arcade)
9	Kuiqi tunnel	1596.1	Fujian Province	Underground interchange with a widest point of 27.42 m

mostly are combustible and even some flammable. Many of the stacked goods are combustible, while the passengers' luggage are placed everywhere, causing difficulty in combustible product control.

Some walls and ceilings of large modern transportation hubs are furnished with wooden or polymer materials for visual effects, which increases the fire load of the stations. And when in flames, the materials will produce large amounts of toxic gases and pose a serious threat to the personal and property safety of the passengers.

(2) Various fire factors

Major modern transportation hubs are developing into bigger scales and more functions and attracting business centers and complicated function areas. First, a large number of electrical devices were installed in different functional areas, such as illumination and decorative lighting, radio systems, air-conditioning systems and some office equipment, and so on. The complicated cables of the electrical devices, some of which are high-power and installed out of sight, or running 24 h a day, any negligence in routine maintenance may cause fire. Second, passengers in the stations are large numbers and with a great mobility and varied consciousness. Some of them may litter cigarette butts or even carry flammable and explosive articles. Sometimes it's difficult for the station staff to spot the hidden dangers and take management measures. Also, the cooking equipment of diners in the transportation hubs further increases the fire risk. In addition, some transportation hubs start operation when still under construction, and construction works are one of the important causes of fires.

(3) Fast spread and destruction of buildings

For the convenience of walking, most areas are vast, interconnected, and well-ventilated. In case of fire, due to the use of decoration materials around stations, it will spread rapidly and burn fiercely. If effective control fails in the early stage of a fire, it will later generate much high-temperature fume and a sharp rise in temperature, and heat radiation will ignite combustibles and expand burning areas. In a large-scale fire, the temperature will maintain at over 1000 Celsius degrees, causing deformation or even collapse of reinforced concrete structures in the station.

Major transportation hubs generally have an atrium. The atrium is linked up with the floor by escalators, and some atriums are even designed as vertical. The structural design is contradictory with fire zoning, creating conditions for the rapid spread of fire and a big fire hazard. In the event of fire, the chimney effect of the atrium will enable quick spread of the smoke to the whole building.

(4) Toxic and harmful substance in fire

The amount of smoke produced by combustibles is related to their physical and chemical properties and combustion states. Cotton and wool clothing, plastics, polymer materials, paper, and wood materials are hard to reach full combustion after being ignited and may release large amounts of toxic gases and soot particles. The smoke will diminish the visibility of the architectural space and hinder the correct identification of escape routes. Some passengers may lose part or all escape capabilities because of poisonous and harmful gases. This is a tremendous threat to the safety of the passengers.

(5) Fire-fighting problems

The area of a large modern transportation hub's single floor is vast and deep. Complicated by the complex construction and the many combustibles, it's difficult to put out a fire from outside. Current fire protection codes generally require large transportation hubs to set valid automatic smoke detectors, automatic alarms, and automatic spray system for the sake of self-help and self-rescue. But due to staff's lack of fire control awareness, lax management, and maintenance of fire-fighting facilities in some transportation hubs, these equipment fail to work during fire accidents. When firefighters arrive, limited by the complex building, malfunctioning fire-fighting equipment, and dangerous environment, they can't effectively reach the ignition source. On the other hand, the opposite movements of passengers and firefighters make it difficult for both sides to move. Therefore, fire in large transportation hubs is a big problem for firefighters.

(6) Difficulties in passenger evacuation

Large transportation hubs have a complex inner space and various passages. They will become mazes for most passengers who are not familiar with the building environment if the signs of guidance are unreasonably set. In case of fire incidents, complex spatial passages will make visitors disoriented and deter their escape. On the other hand, major transportation hub buildings have a huge flow, and passengers are in different states and locations when caught in a fire accident. Therefore, to evacuate such a big crowd dispersed in a vast and complex space within the shortest

possible time is a big challenge, which is aggravated by the passengers in the waiting rooms upstairs. Due to the limited width of the evacuation channel, it's a big
trial to make sure passengers' safe evacuation in an accident.

1.5 Significance of the Research

At present, because of their special function, structure, space, and planning, the
special structural designs of major transportation hubs make the research on new
ways and strategies of fire control of great significance.

As a new method of fire safety design, performance-based fire safety design is
based on fire safety engineering and makes quantitative assessment of a building's
fire risk through its function, structure, materials, and characteristics of the inflam-
mables. Targeting at the complexity and specificity of major transportation hubs,
performance-based design method optimizes the fire control designs basing on the
results of the quantitative assessment, makes it more flexible and cost-effective,
finds the right fire protection measures, and provides an irreplaceable technical and
security support for the construction of large transportation hubs.

In recent years, major transportation hub construction quickened pace, and all
over the world, all large transportation hubs in major cities have completed the planning and been put into operation. Therefore, carrying out studies on performance-
based fire safety design for major transportation hubs is the pressing scientific needs
in urban construction.

Focusing on the fire control strategies for large transportation hubs, which is a
difficulty and hot topic in fire-fighting, this book proposes a performance-based fire
safety design and summarizes commonly applicable fire control designs for large
transportation hubs after application studies on engineering cases.

Chapter 2
The Characteristics of Fire Control in Large Transportation Hubs

2.1 The Functional Areas in Large Transportation Hubs

2.1.1 Ticket Hall

Take Xi' An Xianyang International Airport, for example. The main building of T3A is a very vast ticket hall, and the 327 × 108 m rectangle is supported by 40 columns made with concrete-filled steel tubes, while the fork-shaped steel column supports 20 steel tube trusses above the 35,000 sqm comprehensive hall. The check-in counters in the departure hall abandon the old "room within a room" model, which improves the transparency of the whole comprehensive hall.

Modern, concise, open, vast, clear, and bright, the interior design of Terminal 3A paid special attention to a sense of integrity and orderliness and local cultures, reflecting the distinctive cultural contents and regional features. The architectural elements are fascinating and annotate the character of the space. The streamlined inverted V-shaped primary truss imitates the usage of the space, creating a vibrant and relaxing ambience and signifying the start of a journey. With shortened glass walls, the energy consumption is lowered. The ticket hall has three shared atriums, respectively, on the east and west, which connect with the ground hall and compensate for inadequate daylight (Fig. 2.1).

2.1.2 Waiting Hall

Taking Beijing South Railway Station, for example, the area of its main waiting space is 7990.66 sqm, which is distributed on the ground floor and elevated floor. And the waiting room on the elevated floor has five separate waiting areas, accounting for 7572.5 sqm of the total area, while the VIP room on the ground floor has an area of 418.08 sqm. The table lists the detailed area information of the waiting zones [4] (Table 2.1):

© Springer Nature Switzerland AG 2021
F. Li, H. Li, *Fire Protection Engineering Applications for Large Transportation Systems in China*, https://doi.org/10.1007/978-3-030-58369-9_2

Fig. 2.1 Ticket hall

Table 2.1 Detailed area information of the waiting zones

		Areas of single items		Total areas	
Entrance hall	East(drop-off)		1012.60 m²	4956.90 m²	
	West(drop-off)		1012.60 m²		
	South(entrance)		1378.50 m²		
	North(exit)		1553.00 m²		
Service center	East and west service centers(four)		212.63 m²	850.52 m²	
Ticket zones	Ticket halls(four)		258.65 m²	1034.06 m²	1900.00 m²
	Ticket offices(four)		216.35 m²	865.40 m²	
Waiting zones	Interurban waiting zones	General waiting zones	1955.10 m²	2043.00 m²	
		Soft seat waiting zones	87.90 m²		
	High-speed waiting zones	General waiting zones	2861.61 m²	3311.00 m²	
		Soft seat waiting zones	219.79 m²		
		Group waiting zones	74.40 m²		
		Maternity waiting zones	40.08 m²		
		Soldier waiting zones	40.72 m²		
		Barrier-free waiting zones	74.40 m²		
	General waiting zones	General waiting zones(three)	2166.64 m²	2218.58	
		Soft seat waiting	51.44 m²		
	VIP waiting rooms(two on ground floor)	209.04 m² (per zone)	418.08 m²		
Total area of waiting space	7990.66 m²				

Fig. 2.2 The flexible partition method adopted by the elevated waiting area in Beijing South Railway Station

In terms of partition of functions, apart from the independent soft seat waiting rooms on the south and north, four zones of specific functions were formed using flexible methods, namely, group waiting areas, maternity waiting areas, soldier waiting area, and barrier-free waiting area, as showed by Fig. 2.2.

2.1.3   Boarding Area

A boarding area in the terminal is a place for passengers to wait for boarding and receive corresponding services after security check.

There are altogether four waiting piers in the first phase of the Guangzhou New Baiyun International Airport Terminal, and the third floor of the first pier on the East is for international departures, while the second pier on the east, the first one on the west, and the third floor of the second one on the west are for domestic departures. The waiting hall of the pier is a rectangular with a width of 34 m, and the length of the east and west piers is 345.5 m, the second one on the east and the second one on the west 239 m. The exterior walls are constructed by dotted glass screens and the arch-shaped steel trusses appear every 3 m created a well-organized rhythm. The outdoor view is broad, where you can see the arcs of connected building one of the piers and aircrafts coming and going on the tarmacs (Fig. 2.3).

2.1.4   Baggage Claim

Pudong Airport T1 has a total of 13 baggage carousels, each with a 70 m long part facing the passengers, and 4 of them are used for domestic arrivals, 7 for international arrivals, and 2 for transit flights. For large carrier models with more luggage, the luggage carried by them will be assigned to a specific carousel; several small-scale flights with fewer passengers and less luggage will share a carousel (Fig. 2.4).

Fig. 2.3 Baiyun Airport Pier

Fig. 2.4 Baggage claim hall in Shanghai Pudong International Airport

2.1.5   Arrival Hall

Arrival hall provides arrival procedure services and have arrival corridor, baggage claim, and greeting areas.

The ceiling of the arrival hall and the ceiling of second floor corridor of Pudong International Airport T1 are connected. Apart from the white down lamps on the ceiling, there are also luminescent lamps at baggage claim, which help with the lightning and the direction guidance. Light tanks above the hall columns and on the wall make the place brighter. One side of the entrance hall is covered with aluminum plates in light gray and intermittent advertisings. The height is rather low with ceiling made by white perforated metal plates and the floor by granite (Fig. 2.5).

Fig. 2.5 The interior of Pudong International Airport T1 arrival hall

2.1.6   Transfer Channel

Transfer channels play a very important role by linking the exit areas of high-speed railways with the subway stations or airport terminals. They are indispensable for transfer passengers. But when the crowd is large, narrow transfer channels tend to become a bottleneck point. So in designing a transfer channel, the long-term maximum flow must be taken into consideration. Shanghai Hongqiao transportation hub has a transfer channel on Floor B1, through which passengers can change to maglev, airport terminals and subways, and transfer channels often have information desks (Fig.2.6).

2.1.7   Urban Access Tunnel

With the rapid construction of urban access tunnels, new challenges were brought to urban traffic safety. On one hand, the tunnels are semi-enclosed space; in the event of accidents, they will come jammed. And it's difficult to disperse the traffic in a limited space, especially when it was caught on fire and the semi-enclosed space will make it spread quickly. With the heat and the smog, the drivers will panic, which will lead to a paralyzed traffic system. In this sense, fire control of city access tunnels is a priority as well as a difficulty for designing transportation hubs.

The Shanghai Yangtze River Shield Tunnel has a diameter of 15 m and an internal diameter of 13.7 m. And the shield tunnel segment is 7.47 km in length with the largest diameter in the world. It is also one of the longest underwater tunnels in the world. Within the tunnel, the upper part is for highway and lower part tracks and refuge passages (Fig. 2.7).

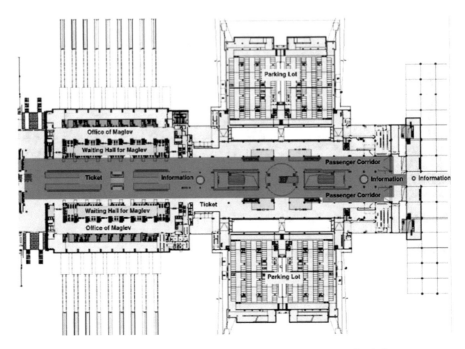

Fig. 2.6 Sketch map of transfer channels in Shanghai Hongqiao transportation hub

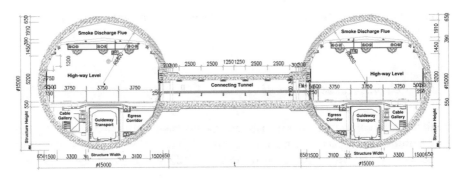

Fig. 2.7 Standard sectional view of Yangtze River Tunnel-shared segment

2.2   Typical Spaces of Large Transportation Hubs

2.2.1   Large and Tall Spaces

Within a large transportation hub, check-in hall, waiting hall, and baggage claim hall normally take up a large space. Construction in a large and tall space focuses on "construction." While "large and tall space" is a specific site, the use of space is of definitive significance. At present, there is still no uniform and accurate definition

Fig. 2.8 An example of large and tall space in a waiting room

for "large and tall spaces" in China or abroad. Neither did NFPA 92B (formulated by National Fire Protection Association), the Guide to Understanding Smoke Control Systems of hall, atrium, and large and tall spaces make a clear definition for large and tall space. Large transportation hub buildings have a large flow, and for the rationalized flow line, they generally are large and tall in design. This design has the following characteristics in term of fire control:

(1) One of the characteristics of large and high buildings is they are high and prone to thermal stratification. The height of the space has a big impact on the patterns and conditions of thermal stratification, and as the height increases, the temperature stratification becomes more prominent. To take full advantage of the characteristic of large and tall building is of great significance in designing smoke exhaust system.

(2) In most buildings with large and tall spaces, the lighting systems concentrate on the canopy. In addition, due to thermal stratification in tall and large buildings, the upper load is heavy, which will lead to two consequences: one, thermal radiation levels at the upper space is very high and easily cause electrical fires, and the other, the accumulated heat will hinder smoke dissipation (Fig. 2.8).

2.2.2 Vertical Space in Atrium

In different periods, the atrium spaces of transportation hubs in different countries have their own characteristics, but they share a common role, that is, the entrance space enclosed on three sides with one side open. Mostly they are linked three-layer spaces. The difference lies in the relationships of the atrium spaces with waiting areas and other functional areas, which affects fire prevention and evacuation strategies. Based on this point, large transportation hubs at home and abroad can be divided into three categories:

(a) Tianjin Railway Station with the renovated shared atrium space

(b) Shenyang North Railway Station with the renovated south station house atrium

Fig. 2.9 Application example of atriums in the buildings of large transportation hubs. (**a**) Tianjin Railway Station with the renovated shared atrium space. (**b**) Shenyang North Railway Station with the renovated south station house atrium

(1) Gallery-style

After China's reform and opening up, during the improvement works, railway stations began to have relatively independent atrium spaces, which were surrounded by commercial spaces. These are gallery-style atriums and the fire control zones are independently partitioned (Fig. 2.9).

(2) Interconnected type

Nowadays in China, when constructing new large-scale transportation hubs, the atrium space of the entrance floor is generally integrated and devoid of shops and can be called shared architectural space in a strict sense. In terms of fire control and smoke control, atrium spaces and elevated waiting rooms can be seen as a whole in order to share fire detection and firefighting equipment (Figs. 2.10 and 2.11).

(3) Hybrid type

The atriums at the entrance floor are often the core space connecting several functional zones with three sides enclosed. They are set with entrance facilities, plenty of commercial space and passages leading to the station. They are similar to the atriums in commercial complexes, and the fire control design for them is similar to those for atriums in commercial complexes (Fig. 2.12).

2.2.3 Underground Space

To meet the needs of urban transportation and other modes of transportation, major transportation hubs generally have underground spaces for transferring to subways, taxis, and buses, and one of the most typical forms of connection is subway system. Subway stations are very important for the entire subway system with most concentrated passengers. As of 2019, Beijing subway daily passengers totaled ten million, more than nine million in Moscow, and eight million in Tokyo [11]. Also, subway

| Atrium of Shenyang north station | Atrium of Nanjing south station | Atrium of Shenyang south station |

Fig. 2.10 Application examples of interconnected atriums. (**a**) Atrium of Shenyang north station. (**b**) Atrium of Nanjing south station. (**c**) Atrium of Shenyang south station

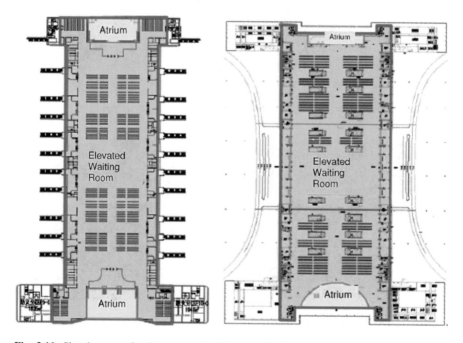

Fig. 2.11 Sketch maps of atrium spaces in Shenyang South Railway Station and Tianjin West Railway Station

stations assemble all the main facilities and operating systems, and fire control in subway stations is a top priority in firefighting engineering and management. Subway stations can be divided into three parts: the platform and station hall, entrance and exit and corridors, and ventilation system [12].

The platforms and station halls form a long and narrow space, where people make the most movements; entrance and exit and corridors are for passengers to enter and leave the stations, but, in case of fire, they are the passage for fresh air; the

Atrium of Yokohama Railway Station, Japan   Atrium of Tokyo Railway Station, Japan   Atrium of Kyoto Railway Station, Japan

Atrium of Luzern Railway Station, Switzerland   Atrium of Berlin Railway Station, Germany

Fig. 2.12 Application example of interconnected atriums in station buildings

Fig. 2.13 Subway station transfer scenarios

ventilation system replenish air mechanically to keep good air quality, and, in case of fire, to dissipate smokes mechanically. At present, the subway networks are well-developed in some large cities, and some subway stations even have adopted multi-layer structures. In addition, in order to make use of space, stores were even opened at some subway stations, which has increased the subway stations' fire load and posed new problems and challenges to fire security (Fig. 2.13).

Subway stations are typical long and narrow underground spaces, which is characterized by the following: first, the space is relatively narrow. In case of fire, smoke will spread toward two ends and accumulate. With the presence of spontaneous combustion and mechanical air replenishment, the fire will possibly expand further;

second, with a great fuel load, fire will spread fast and become more severe; third, because the crowd is concentrated and dense, in the event of flashover, it may cause mass casualty or mass injury (Table 2.2).

1. Characteristics of underground space firefighting

2. The evacuation problems in underground spaces
   The difficulties of evacuating people from underground spaces during fire (Table 2.3):

2.2.4   The Long and Narrow Space of Tunnels

Fire in urban traffic tunnels can be caused by many factors, which are characterized by:

(1) Moving igniter
   Subway trains run through the tunnel continuously, and the uncertainty of a fire determines that the train can catch fire at any position in the tunnel. This will bring difficulties to alarming systems, timely evacuation of people and fire extinguishing. In order to facilitate the timely response and emergency rescue after ringing the alarm, the subway trains should maintain traction power, drive away from the tunnel, and reach the reserved open outdoor area for complete dealing. So, mobility is a remarkable feature of subway tunnel fire.

(2) Varied fire causes
   Subways have very centralized mechanical electrification with lots of electrical equipment, which increases the possibility of electric fire. In addition, they are driven by mechanical energy, which poses the danger of overheating and fire by friction. Passengers may bring high-risk items such as lighters into the station due to security negligence, whose improper storage may cause fire.

Table 2.2 The combustion characteristics of fire in underground buildings

	Characteristics
Combustion condition	In a sense, the combustion conditions of underground buildings are determined by the external ventilation. Due to limited numbers of entrances and insufficient oxygen supply, incomplete combustion will happen, causing thick smoke, which gradually spread and dissipate at the entrances
	Fresh air passes though the entrances to reach the underground buildings and form a neutral plane, which is high at the beginning of fire and gradually reduces
Mobility delay	The flow of smokes and gases in underground building is very complex and changes under the influence of the wind direction and wind speed on the ground, especially for underground buildings with two or more entrances, where air inlet and smoke outlet are naturally formed. The more entrances, the faster the fire burns

Table 2.3 The characteristics of evacuating people from underground spaces

Type	Characteristics
The limitation of extinguishing methods	Restricted by conditions, underground buildings have limited entrances, and the walking distance is long during evacuation. In the event of fire, people can only be evacuated through the entrances, while fire ladders and other rescue tools are useless for underground building evacuation
One-way evacuation	The evacuation during fire in an aboveground building is two-way. People on the burning floor can be evacuated downward or upward to the rooftop. However, when the underground buildings catch fire, the evacuation is only upward and people won't be safe until they get out of the underground buildings
From entrances and exits to smoke outlet	During a fire, the entrances and exits will become smoke outlets in the absence of smoke exhausting facilities. The diffusion of high-temperature smokes and the evacuation of the people are in the same direction, and underground buildings are more dangerous. Researches in China and abroad confirm that vertical speed of smoke is 3 ~ 4 m/s and its horizontal diffusion rate is 0.5 ~ 0.8 m
The panic of the crowd	There are no natural lights in the underground spaces, and power failure and the automatic cutoff of fire-fighting systems will make people disoriented and panic. Plus their unfamiliarity of organization and evacuation routes in the underground spaces, their fear and sense of danger will increase after the delayed evacuation
Difficulty in moving upward	During the evacuation in underground buildings, people have to walk up instead of walking down the stairs, which is much more difficult than the evacuation in aboveground buildings, thus slowing down the speed. The direction of evacuation and that of high-temperature smoke proliferation is the same, bringing great difficulty for evacuation
Death caused by hypoxia	Hypoxia is common in a burning underground building filled with large amounts of carbon monoxide and other harmful gases, which are extremely harmful to humans. Hypoxia is more severe in burning underground buildings than in ground buildings. According to fire casualty statistics, the casualty caused by hypoxia account for 85% of the total death in burning underground buildings

(3) The flow of smoke and gas in closed spaces

Urban underground railway tunnels are usually long or extra-long. Without compulsory ventilation, in such an enclosed space, the controlled combustion of fuel won't last long. The combustion is primarily determined by ventilation, and incomplete combustion will produce carbon monoxide and other hazardous substances.

At the early stage of fire, as a result of thermal buoyancy, level wind pressure, and piston wind effect, the smoke and gases will flow in the airtight and narrow space. With the development of the fire, gas and smokes gradually go down and disperse along the cross sections of the underground tunnels.

(4) Difficulties of evacuation

Subway tunnels are underground space buildings with a great buried depth, and the evacuation to the ground takes a longer time. In addition, underground spaces fully rely on artificial lighting. Once fire causes a power outage; it is bound to cause alarms and chaos. And the tunnels are also for the spread of smoke, gases, and combustion. Restricted by heat and toxic gases, the evacuation ability will be undermined, and this will make it difficult to evacuate.

2.3 Fire Risk of Typical Components in Major Transportation Hubs

2.3.1 Fire Risk of Steel Structure

The unique features and needs of large transportation hubs not only promote the application of large-span spatial structures but also create a structural system integrating bridges and buildings, which enriched the spatial structure. With the advancement of architectural modeling of large transportation hubs, the structural design also takes on new looks, and transportation hubs become the perfect combination of structures and architectural shapes. For a major transportation hub, the most important space is the elevated waiting rooms and truss of the terminal. Depending on the architectural appearance and functionality, the roofs of long-span spaces are built using steel truss or steel cable systems, with steel being the main material. Table 2.4 summarizes the steel application in large transportation hubs.

Steel materials are more preferable compared with reinforced concrete, allowing the implementation of super large space conceptions. But the fatal flaw of steel structures is their inability to withstand high temperatures. So the most steel roofs are treated with fire retardant material. Steel grid structures used by major transportation hubs have the following characteristics:

1. The characteristics of steel structures used by large transportation hub station
   Most major transportation hubs used steel roof to realize large-span waiting rooms. According to the division of fire protection zones, performance-based fire control design use steel structures in open spaces or large spaces (Fig. 2.14).

2. High-temperature property of steel structures
   The performance of bare steel structure systems in a fire will drop due to temperature rise and the damage caused, as shown by Table 2.5 [5]

3. Fire endurance
   The confirmation of fire endurance has to be in compliance with relevant provisions. When designing multifunctional large-span steel structures, fire scene simulations based on scientific methods must be done in order to find reasonable protection schemes.

   Railway transportation hubs have a tall and vast space and a low fire load density, and in case of fire, the temperature rise of railway transportation hubs is differently from small room. So the fluid dynamics analysis of fire is adopted to get the temperature curve and design fire control structures.

Table 2.4 Application example of steel in large transportation hubs

Typical station	Application of steel truss	View
Wuhan Railway Station	Cover-up double layer grid shell two-way one-piece truss. The station hall and the rain shelter of the platform share a roof with a span length of 116 in the center	

Typical station	Application of steel truss	View
Nanjing South Railway Station	Steel structured roof; the truss system has a north-to-south length of about 450 m and an east-to-west length of about 200 m; the total area is about 100,000 sqm	
Beijing South Railway Station	Roof truss is hyperboloid oval-shaped, with a long axis of 350 m and a short axis of 190 m, the main steel frame under force is side-span steel trusses and solid box girder	

(continued)

Table 2.4 (continued)

Typical station	Application of steel truss	View
Beijing Daxing International Airport	Steel structure of the roof truss has an irregular surface, spanning up to 180 m, with a 30 m disparity between the highest and the lowest points	

Typical station	Application of steel truss	View
Beijing Capital International Airport	The roof of Beijing Capital Airport T3B terminal is a 958 × 775 m long-span hyperboloid cutout broach mixed with node grid structures, the truss area is 141,000 sqm	

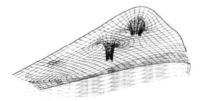

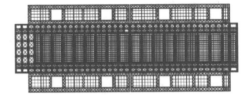

(a) The model of a corner at Beijing   (b) The model of Shanghai Hongqiao
Daxing Airport built by steel grid   Station's steel roof

Fig. 2.14 The model of steel roof at large transportation hub. (**a**) The model of a corner at Beijing Daxing Airport built by steel grid. (**b**) The model of Shanghai Hongqiao Station's steel roof

Table 2.5 Changes in steel strength at different temperatures

Temperature/°C	Performance of steel
<175	Slightly rises after being heated but drops significantly as temperature rises
500	Only 30% of that in normal temperature
750	Completely lost

4. Fire protection difficulties of steel structure and general methods for fire control
(1) Fire control scheme

According to engineering cases, thick or thin fire retardants are commonly used for the protection of steel structures; Table 2.6 shows the main methods [6]:

(2) Fireproof protection layer
Tables 2.7 and 2.8 shows construction methods:

2.3.2   The Fire Hazards of Glass Wall

Glass wall is a type of new contemporary wall whose most prominent features are the integration of architectural aesthetics, function, energy saving, and structures, and it enables the building to present different beauties from different perspectives, such as the dynamic beauties brought by sunlight, moonlight, and lamplights. To fully reflect the characteristics of large-span and high spaces in contemporary major transportation hubs, meeting the requirements of noise, energy efficiency, thermal engineering, and main enclosing part will feature extensive use of glass curtain wall. For example, Wuhan Railway Station and terminals at Pudong International Airport heavily uses glass wall, and the vertical glass wall of the former reaches a staggering 30,000 sqm (Fig. 2.15).

Table 2.6 The characteristics and application scopes of fire protection methods for steel components

No.	Method		Characteristic and scope of application	
1	Covered with concrete or masonry		High-strength, impact resist, take up more space, difficult to use on steel beams and inclined constructions, suitable for fire control steel columns prone to collision and without supporting panels	
2	Coated with fire retardant	Intumescent (thin)	Lightweight, simple to use, suitable for any shapes and components, mature technology, wide application, but strict on coating substrates and environmental conditions	When fire endurance <1.5 h, exposed steel structure with decoration requirements
		Non-intumescent (thick)		Good durability, good fire protection
3	Fireproof coating		Good prefabricate quality, high integrity, stable performance, smooth surface, clean and bright, good decoration, without construction limitations, especially suitable for cross operation and non-wet construction	
4	Covered with flexible blanket-like insulation materials		Good heat insulation, easy construction, low cost, suitable for indoor areas less vulnerable to mechanical damages and wetness	
5	Composite fire protection	2b + 3	Good heat insulation and integrity, decorative, suitable for steel column and steel beam with high decoration requirements and high fire performance requirements	
		2b + 4		

Table 2.7 Fire protection layer usually adopted in steel columns, beams, and trusses

Construction method	Site operation		Prefabricate		
	Pour	Spray	Plate	Irregular plate	Blanket
Shape	H-shaped		H-shaped or box-shaped	Box-shaped	
Steel column Steel plain web beam Steel truss	Very applicable Rather applicable Very applicable	Rather applicable Very applicable Very applicable	Very applicable Rather applicable —	Very applicable Very applicable —	
Material	Lime concrete	Jetted concrete vermiculite Mortar Mineral fiber mortar Perlite mortar Perlite and vermiculite plaster	Plaster tablet Plaster board Asbestos silicate sheets Fiber silicate board Vermiculite silicate board	Plaster pieces Perlite piece Silicate piece	Mineral fiber Blanket

Table 2.8 The construction methods of fire protection layer for steel

Construction technology	Construction method
Casting	The most reliable method of fire control for steel structure generally uses concrete, lightweight concrete or air-entrapping concrete
Spraying	Can be divided into two methods, one, direct spraying, and, the other, weld steel mesh on H-shaped steel structure and spray fire protection material to form a hollow layer. Spray materials are normally insulation materials such as rock wool and slag wool, currently the most used method of fire protection for steel structure
Suspended ceiling	Use light, thin, fire-resistant materials to build fire-controlling ceilings and save fire protection layers such as steel frame, steal truss, and steel roofs
Pasting	Make prefabricated board using fire protection materials such as asbestos calcium silicate, mineral wool, and light gypsum board, and attach it to steel structure with bonding agent; if bolts, rivets, and other components on the binding area are not smooth, first paste interlay boards on them, and then paste the protective sheet on the interlay boards
Combination	Combine two or more fire protection material to control fire

Fig. 2.15 Application examples of glass walls in Wuhan Railway Station and Pudong International Airport Terminals

Wuhan Railway Station is located in Hongshan District of Wuhan, Hubei Province. Adjacent to the Third Ring of Wuhan, it is an important station on Beijing-Guangzhou high-speed railway and China's first high-speed train station, as well as one of Asia's largest high-speed railway stations. The station facades are made of glass walls, which altogether have a total area of 30,000 sqm and 12,000 pieces. To highlight the city's identity as a lake capital, about 5000 pieces of irregular glasses were used to make a wave shape. The intensive glasses impart a modern and transparent feel to the station, which won the International Architecture Award issued by Athena Design Museum in Chicago, USA.

However, because of glass walls' special physical properties, they have relatively low fire-resistant performances, and extensive use of glass walls causes certain difficulties in the building's fire control; Table 2.9 lists the difficulties in fire control brought by glass walls:

Table 2.9 Difficulties in fire control brought by glass walls

Definition		Glass wall is the exterior structure of building made up of metal components and glass
Characteristics		Lightweight, bright, firm, beautiful, decorative
Difficulty of fire control	Broken glasses	In case of fire, the room temperature of the buildings will rise rapidly in the initial stage, and large areas of glass wall will burst because of temperature stress, causing rapid spread of fire and making an "ignition channel." Fire will spread to the upper or adjacent partitions and ignited the inflammables in them
	Enlarging gaps	Gaps between vertical glass walls and horizontal slabs become the channel for spreading the fire. Due to construction requirements, a gap is between the glass wall and the floor, forming an ignition duct and destroying the fireproofing partition and making the fire spread to each floor; the higher the floor, the stronger the ignition ability
	Poor thermal insulation	Glass wall with poor thermal insulation which needs to be protected by cooling spray system

2.4 Human Factors

Human factors are the basis of evacuation analysis and the precondition for evacuation safety evaluation. Based on a detailed and accurate analysis of people load, human factors are influenced by countries, regions, and types and functions of a building and restricted by the space layout, the floorage of each room, and the internal layout.

The main passengers of large urban transportation hubs (such as subways and light railways) use them daily and are familiar with their spatial plan. The passengers of a major inter-urban transportation hub (such as an airport terminal, a high-speed trains, or bullet train) normally use them for the first time or with little experience, and they are unfamiliar with the space. During workday traffic peaks, urban transport facilities will be very crowded, while on holidays, the passenger flow will be more even. The number of passengers in railway stations will be the highest on holidays. To fully understand the distribution features of passenger flow will provide references for evacuation plans in fire control designs for large transportation hubs.

2.4.1 The Characteristics of Passenger Flow in Transportation Hubs

Evacuation of people is an essential parameter in fire control design, and analyzing the characteristics of passenger flow is of reference value to determine the number of the most disadvantaged in an emergency evacuation. Compared to air passengers, rail passengers are large in number and with prominent characteristics, and this section is dedicated to analyzing the characteristics of rail passenger flow.

Time distribution of rail passenger flow refers to the daily passenger volume at each time period. In different time span, passenger flow has a different changing pattern. According to the time of measurement, the passenger flows can be divided into workday flow, weekend passenger flow, and holiday passenger flow.

1. The distribution of workday passenger flow [7]

Workday passenger flow is based on the distribution of passenger flow at different periods of time. Daily passenger statistics generally is a statistical indicator for urban rail transportation. Through calculating the hourly passenger flow, daily dispatching plans and train plans are formulated to guide the development of a transport scheme.

Variations of passenger flow are linked with the habits of the people in the region and often subjected to the factors of commuting time, school-going or home-going time of students, directions of the lines, and the network structures. This part will introduce the daily passenger flow into high-speed railways and summarize the distribution characteristics of passenger flow into four categories, including single peak type, double peak type, all peak type, and abrupt peak type, as shown by following figures (Figs. 2.16, 2.17, 2.18, and 2.19).

But in real life, due to the complexity of influence factors, daily passenger flow distribution is not entirely in accordance with the four peak types. Instead, it is featured in the different periods of time by different peak types, which makes the study on daily passenger flow distribution particularly critical.

2. The distribution of weekend passenger flow

In this section, weekend refers to the time beginning from Friday to the end of the weekend, when the passengers have a much higher travel need than the other

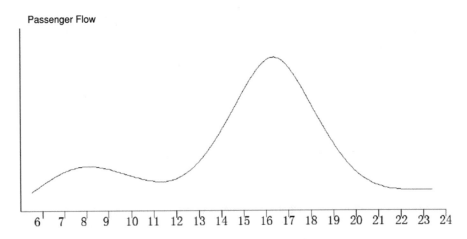

Fig. 2.16 Distribution characteristics of single peak passenger flow

Passenger Flow

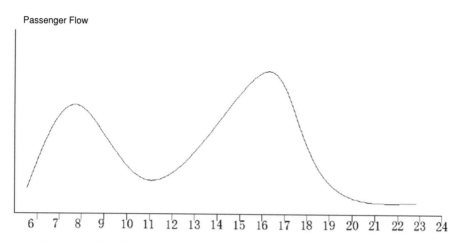

Fig. 2.17 Distribution characteristics of double peak passenger flow

Passenger Flow

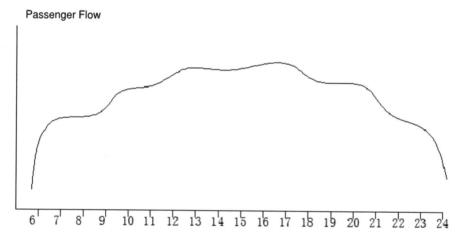

Fig. 2.18 Distribution characteristics of all peak passenger flow

4 days. Due to the changes in travel demands, the characteristics of passenger flow distribution on weekend are different from those on workday.

3. The distribution of holiday passenger flow

Here, holidays refers to New Year, Spring Festival, Tomb-Sweeping Day, Labor Day, National Holiday, and summer and winter holidays. Due to varied durations, passenger types, and regional differences, holiday passenger flows also take on different forms.

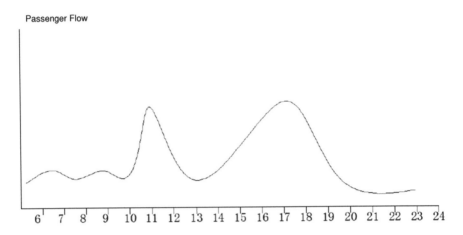

Passenger Flow

6 7 8 9 10 11 12 13 14 15 16 17 18 19 20 21 22 23 24

Fig. 2.19 Distribution characteristics of abrupt peak passenger flow

2.4.2 Effects of High-Density Crowd to Evacuation in Fire

1. Detention clustering

Detention clustering is common in case of fire in major transportation hubs. When the indoor population evacuation density is large ($0.5 < P < 3.8$ persons/sqm), a cluster will be formed, and the individual will solely depend on the collective activity trends.

During security check and check-in, people tend to cluster, while a fire is particularly prone to a cluster effect in the station. Passengers usually stay longer in places like hallway entrances and staircases, and in case of fire, two passenger flows, one in and one out, tend to scramble to escape. The flow velocity is zero in this circumstance, causing failed evacuation. This would significantly delay the evacuation and lead to stampede and serious casualties.

In order to reduce this possibility, evacuation layout should avoid abrupt spatial change and provide several evacuation directions and easily identifiable evacuation signs to help people avoid panic or disorientation [8] (Fig. 2.20).

2. Evacuation density of people

The evacuation density of people influences the evacuation activities in a railway station in terms of order of evacuation and crowd psychology (Table 2.10).

2.4.3 Analysis of Evacuee Number

Due to the complex functions inside large transportation hubs, the people needing evacuation include the station staff and passengers. Station staff's activity areas are relatively fixed, while passengers are of higher mobility.

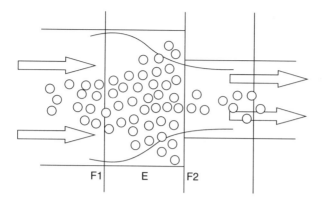

Fig. 2.20 Sketch map of cluster evacuation

The analysis of the number of people to be evacuated is based on the functionalities of different spaces, confirmed by statistical data of the areas used by station staff, while the areas occupied by passengers are analyzed by field survey and sampling and by using a passenger flow formula.

Passenger flow formula:

$$\text{The number of people} = \frac{\text{passenger flow}(\text{Person / hour}) * \text{Duration of stay}(\text{minutes})}{60}$$

Taking Shanghai Hongqiao transportation hub as an example, it is expected to serve 78,380,000 passengers in 2030, and the maximum number of people assembled is around 10,000. The number of station staff can be deduced by the floor area. Based on the assumption that every passenger stays for 40 min in the waiting hall, it can be converted to nearly 5 min; and passengers coming in or leaving will stay at the station hall for about 10 min. Based on the data, increase an amplification coefficient of 30%, the evacuee numbers of the two large regions can be deduced (Tables 2.11, 2.12, and 2.13).

2.4.4   Calculation of Move Time

During the passenger evacuation process, the time taken by stairs, hallways, and the entrance passages decides the evacuation time of the entire space. And the time spent at a region is closely related to crowd density, walking speed, and width of the entrances; the information shown in the table below is based on statistical data (Table 2.14):

An important indicator for movement time is walking speed, which directly determines the overall evacuation time, particularly of large waiting areas or exit areas. Walking speed is determined by many complex factors, including the variation of the evacuees, the distance between people, and the evacuation environment.

Table 2.10 Effect of people density on safe evacuation

Contributory factors	Description of effects
Activity order	Under normal circumstances (that is, a non-accidental event occurs), people tend to gather in an area of no more than 0.28 sqm per capita, which will cause potential dangers. When the per capita area was further reduced to 0.25 sqm, there is physical contact between people. In emergencies, any activity may cause congestion, pushing, fell-over, stampede, and injury and casualty. Therefore, a crowd density of 0.28 sqm per capita is an important parameter for evacuation
Crowd psychology	Unable to understand the disaster and the state of the environment, stranded individuals in the crowd will panic and are prone to irrational behaviors Due to high use rate of major transportation hubs, during rush hour, crowd may gather in ticketing, security, ticket check, and entrance and exit areas. In planning floor layout and fire evacuation strategies, the impact of overly crowded passengers must be taken into consideration

Table 2.11 The head counts of each areas at Shanghai Hongqiao traffic hub

Location floor		Function	Different crowd types	Number	Number (Person)
Elevated mezzanine	16.1 m	Office use	The number of people in rush hours (people per hour)	1047	1047
	21.95 m	Commercial use	The number of people in rush hours (people per hour)	2347	2347
Elevated floor		Waiting rooms for Beijing-Shanghai lines	The number of people in rush hours (people per hour)	5200	
			Seeing off guests(30%)	1560	
			People margin(30%)	2028	
			Duration of stay(minute)	40	5859
		Waiting rooms for inter-urban lines	The number of people in rush hours (people per hour)	1648	
			Seeing off guests(30%)	494	
			People margin(30%)	643	
			Duration of stay(minute)	40	1856
		Commercial service	The number of people in rush hours (people per hour)	921	921

(continued)

Table 2.11 (continued)

Location floor	Function	Different crowd types	Number	Number (Person)
Platform floor	East-west entrance halls	The number of people in rush hours (people per hour)	32,315	
		Duration of stay(minute)	5	2692
	Local police station and station offices	The number of people in rush hours (people per hour)	765	765
	Platforms	Assume each of the 2 trains has 1280 passengers	2560	
		30% pick-up and drop-off guests	766	3328
B1	Exit hall for Beijing-Shanghai lines	The number of people in rush hours (people per hour)	4983	
		Seeing off guests(30%)	1495	
		People margin(30%)	1943	
		Duration of stay(minute)		1404

Table 2.12 Areas taken by different types of rooms(m²/person)

Building type	Room type	Usable floor area per capita
Office building	General office	4
	High-end office	8
	Meeting room	2.5
	Corridor	50
	Others	20
Hotel building	General office	15
	High-end office	30
	Meeting room	2.5
	Corridor	50
	Others	20
Mall building	General stores	3
	High-end store	4

Table 2.13 Head count at Hongqiao transportation hub

Location	Number of people	Remark
11.55 m B1	7869	
0.00 m platform floor	3328	Take into consideration two train arrive at the same time
0.00 m east-west entrance hall	2692	
10.1elevated waiting hall	8636	
Office or commercial mezzanine	3394	
Total number	25,919	

Table 2.14 Entrance flow in congestions

Facilities	Degree of congestion	Density(people/ m²)	Speed(m/ minute)	Number of people passing through 1 m width per minute
Stairs	Least	0.54	45.60	16.45
Stairs	Medium	1.09	36.48	39.50
Stairs	Optimal	2.07	28.88	59.19
Stairs	Badly	3.26	12.16	39.50
Corridor	Least	0.54	76.00	39.50
Corridor	Medium	1.09	60.80	65.79
Corridor	Optimal	2.17	36.48	78.99
Corridor	Badly	3.26	18.24	59.19
Entrance passage	Medium	1.09	51.68	55.95
Entrance passage	Optimal	2.39	36.48	85.48
Entrance passage	Badly	3.26	15.20	49.34

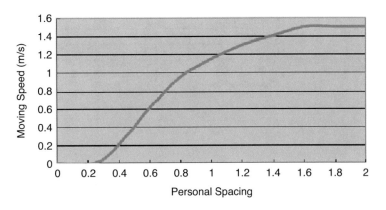

Fig. 2.21 Relation curve of moving speed and personal spacing

Table 2.15 Passenger composition and their walking speeds

Person T\type	Percentage(%)	Speed when walking fast(m/s)		Speed of going upstairs(m/s)	Speed of going downstairs(m/s)
		Fastest	Best		
Male(aged 15–29)	23.0	1.200	0.960	0.536	0.808
Male(aged 30–50)	32.0	1.200	0.960	0.504	0.688
Male(aged 51–80)	2.50	1.200	0.960	0.408	0.536
Female(aged 15–29)	17.0	1.200	0.960	0.508	0.604
Female(aged 30–50)	21.0	1.200	0.960	0.472	0.532
Female(aged 51–80)	2.50	1.200	0.960	0.388	0.388
Children	2.00	1.200	0.960	0.388	0.388

According to statistical data, the following illustration shows the impacts on speed by the distance between people (Fig. 2.21, Table 2.15):

2.4.5   The Composition and Characteristics of People

2.5   Brief Summary

Beginning with the incoming and outgoing flows of passengers at large transportation hubs, this chapter summarized two major flow patterns at large transportation hubs: "enter above and exit below" and "enter below and exit below." Secondly, the chapter summarized the various types of function space at a large transportation hub and, from the angle of fire control, summed up the typical features in major transportation hubs. From the perspective of the passenger flow in the organization, the features of space functions, typical components of construction, and human factors, it analyzed the characteristics of fire control in large transportation hubs. And the result shows all transportation hubs share a common character: a large passenger flow, but the distribution pattern is very different, with subway passenger flow featuring several peaks and the passenger flow of high-speed railway station featuring several models caused by various factors. Looking at the space features, large transportation hubs are characterized by large space, atrium, and narrow transfer corridors. To integrate them into a transportation hub, they have to have their own spatial characteristics and meet the function requirement, which requires wise combination and division of the spaces.

Chapter 3
Fire Safety Design for Large Transportation Hubs

3.1 Development of US and Chinese Standard Systems

3.1.1 History of US Fire Safety Standards for Transportation Hubs

The US standards on fire safety design for transportation hubs are mainly based on the NFPA 130. The history of this standard will be briefly described in this section [9].

The Fixed Guideway Transit Systems Technical Committee was formed in 1975 and immediately began work on the development of NFPA 130. One of the primary concerns of the committee in the preparation of this document centered on the potential for entrapment and injury of large numbers of people who routinely use these types of large transportation hubs.

During the preparation of the first edition of this document, several significant fires occurred in fixed guideway systems, but fortunately the loss of life was limited. The committee noted that the minimal loss of life was due primarily to chance events more than any preconceived plan or the operation of protective systems.

The committee developed material on fire safety requirements to be included in NFPA 130, Standard for Fixed Guideway Transit Systems. This material was adopted by NFPA in 1983. The 1983 edition was partially revised in 1986 to conform with the NFPA Manual of Style. Incorporated revisions included a new Chapter 8; a new Appendix F, Creepage Distance; minor revisions to the first four chapters and to Appendices A, B, C, and E; and a complete revision of Appendix D.

The scope of the 1988 edition was expanded to include automated guideway transit (AGT) systems. The sample calculations in Appendix C were revised, and Appendix D was completely revised.

The 1990 edition included minor changes to integrate provisions and special requirements for AGT systems into the standard. Table 1 from Appendix D was

© Springer Nature Switzerland AG 2021
F. Li, H. Li, *Fire Protection Engineering Applications for Large Transportation Systems in China*, https://doi.org/10.1007/978-3-030-58369-9_3

moved into Chapter 4, Vehicles, and new vehicle risk assessment material was added to Appendix D.

Definitions for enclosed station and open station were added in the 1993 edition, along with minor changes to Chapters 2 and 3; the 1995 edition made minor changes to Chapters 1, 2, and 3.

The 1997 edition included a new chapter on emergency ventilation systems for transit stations and trainways. A new Appendix B addressing ventilation replaced the previous Appendix B, Air Quality Criteria in Emergencies. Also, the first three sections of Chapter 6 (renumbered as Chapter 7 in the 1997 edition), Emergency Procedures, were revised, and several new definitions were added.

The 2000 edition of NFPA 130 addressed passenger rail systems in addition to fixed guideway transit systems. The document was retitled Standard for Fixed Guideway Transit and Passenger Rail Systems to reflect that addition, and changes were made throughout the document to incorporate passenger rail requirements. Additionally, much of Chapter 2 was rewritten to incorporate changes that were made to the egress calculations in NFPA 101 ®, Life Safety Code®. The examples in Appendix C were modified using the new calculation methods. The protection requirements for Chapter 3 were modified, addressing emergency lighting and standpipes. Chapter 4 also was modified to clarify and expand the emergency ventilation requirements.

The 2003 edition was reformatted in accordance with the 2003 Manual of Style for NFPA Technical Committee Documents. Beyond those editorial changes, there were technical revisions to the egress requirements and calculations for stations. The chapter on vehicles was extensively rewritten to include a performance-based design approach to vehicle design as well as changes to the traditional prescriptive-based requirements.

The 2007 edition included revisions affecting station egress calculations, the use of escalators in the means of egress, vehicle interior fire resistance, and power supply to tunnel ventilation systems. The chapter on vehicle maintenance buildings was removed because requirements for that occupancy are addressed in other codes; the performance-based vehicle design requirements were substantially revised to more accurately address the unique qualities of rail vehicles.

The 2010 edition of NFPA 130 included provisions that allowed elevators to be counted as contributing to the means of egress in stations. The 2010 edition also contained revisions relating to escalators, doors, gates, and turnstile-type fare equipment. The units in the standard were updated in accordance with the 2004 Manual of Style for NFPA Technical Committee Documents. Several fire scenarios were added to Annex A to provide guidance on the types of fires that can occur in vehicles, stations, and the operating environment as well.

The 2014 edition of NFPA 130 included substantial reorganization of Chapters 5 and 6 for consistency and consolidation of wire and cable requirements into a new Chapter 12. Other changes included reconciliation of terminology related to enclosed trainways and engineering versus fire hazard analyses, revisions to interior finish requirements, revisions to requirements for prevention of flammable and combustible liquid intrusion in Chapters 5 and 6, and improvements to Annex C.

After 2017, the latest edition of NFPA 130 has added several new definitions and modified requirements for materials used as interior wall and ceiling finishes. Enclosed stations are now required to be equipped with a fire alarm system and stations, and enclosed trainways are now required to be equipped with an emergency communication system, as outlined in revised Chapter 10. A new Annex B now provides guidance on establishing noise levels in order to maintain a minimum level of speech intelligibility through the emergency communication system. In Annex C, modifications have been made to the example showing means of egress calculation. A new Annex H provides information on fire scenarios and methodologies used for predicting fire profiles.

3.1.2 History of Chinese Fire Safety Standards for Transportation Hubs

The Chinese standard system of fire safety design for buildings is based on the Fire Code for Architectural Design, which is approved by the Ministry of Housing and Urban-Rural Development of the People's Republic of China and has legal effect on most buildings. Additionally, China has developed standards on fire safety design for large transportation hubs including the Fire Code for Metro Design, the Fire Code for Railway Engineering, and the Fire Code for Civil Airport Terminal Design. As these three standards only specify fire safety requirements for areas specific to large transportation hubs and the fire safety design for other areas should be executed in accordance with the Fire Code for Architectural Design, the author will briefly describe the development of all these four standards in a chronological order.

In April 1956, the 102-56 Interim Fire Safety Standard for Industrial Corporate and Residential Buildings, formulated by the Ministry of Public Security and the Ministry of Construction Engineering, was examined and approved by the National Construction Commission; its trial implementation began on September 1, 1956. This standard, which was based on the Soviet standard H102-54, contained basically the same chapters with amendments and supplements made to some clauses according to China's realities of that time. It was the first fire safety standard for architectural design in New China.

In 1960, the Regulation on Fire Safety Principles in Architectural Design was jointly promulgated by the National Infrastructure Commission and the Ministry of Public Security. This regulation, which included only 8 articles and more than 1000 characters, was not a standard, but had legal effect. In order to facilitate the implementation of this regulation, the aforesaid commission and ministry formulated the Technical Data on Fire Safety in Architectural Design for reference by design departments. The Technical Data on Fire Safety in Architectural Design, which contained basically the same chapters as the standard 102-56 with slight amendments to it, had no legal effect and was only an information annex to the Regulation on Fire Safety Principles in Architectural Design. The introduction of the regulation shows that the standard 102-56 wasn't strictly implemented and that relevant national departments must reiterate the important provisions on fire safety in architectural design.

In 1975, the TJ16-74 Fire Code for Architectural Design, approved by the National Infrastructure Commission, the Ministry of Public Security, and the Ministry of Fuel Chemical Industry, was issued in China; its trial implementation began on March 1, 1975. The code, which included nine chapters and six appendices, was revised based on the Regulation on Fire Safety Principles in Architectural Design.

Since the reform and opening-up policy was implemented, China's economy has developed rapidly, and a large number of originally rarely seen high-rise buildings have sprung up across the country. In order to ensure the fire safety of high-rise buildings, the provisions on fire safety of high-rise buildings, which were not included in the TJ16-74 Fire Code for Architectural Design, were urgently needed. But on the other hand, as the TJ16-74 code took effect just a few years ago, on March 1, 1975, only a few revisions were needed. Therefore, the Chinese government decided to develop and implement a separate fire safety standard for high-rise civil buildings.

In 1982, the GBJ45-82 Fire Code for High-Rise Civil Architectural Design (Trial Implementation), prepared by the Ministry of Public Security, was jointly approved by the National Economy Commission and the Ministry of Public Security; its trial implementation began on June 1, 1983.

On May 13, 1995, a new edition (GB50045-95) of the Fire Code for High-rise Civil Architectural Design, revised by the Ministry of Public Security in conjunction with relevant departments, was examined by relevant ministries and approved by the Ministry of Construction as a compulsory national standard; the code came into effect on November 1, 1995. It was partially revised in 1997, 2001, and 2005.

Following the introduction of the Fire Code for High-Rise Civil Architectural Design, the Fire Code for Architectural Design was also revised. On August 26, 1987, the GBJ16-87 Fire Code for Architectural Design, revised by the Ministry of Public Security in conjunction with relevant departments, was examined by relevant ministries and issued as a national standard. The code came into effect on May 1, 1988. It was partially revised in 1995, 1997, and 2001 and had 19 chapters and 5 appendices.

On October 25, 1999, the TB10063-99 Fire Code for Railway Engineering was developed and approved by the Ministry of Railways of the People's Republic of China; the code took effect on February 1, 2000. It was the first edition of the Fire Code for Railway Engineering in China.

On July 12, 2006, the GB50016-2006 Fire Code for Architectural Design was introduced by the Ministry of Construction and the General Administration of Quality Supervision, Inspection and Quarantine; it took effect on December 1, 2006.

With the rapid economic and social development in China, the GB50016 Fire Code for Architectural Design (2006) and the GB50045-95 Fire Code for High-Rise Civil Architectural Design (2005) had been unable to meet the practical engineering needs. All kinds of fire accidents also reflected some problems that needed urgent improvement. In addition, some inconsistencies between the two standards and the national engineering standard system and other fire safety design codes posed great difficulties to the implementation of these codes.

On December 29, 2007, a revised edition of the Fire Code for Railway Engineering was approved and implemented by the Ministry of Railways of the People's Republic of China. This edition analyzed the latest fire safety practices in railway engineering with reference to the national fire safety standards. It was partially revised in 2012.

In this context, the newly developed Fire Code for Architectural Design was approved on August 27, 2014 by the Ministry of Housing and Urban-Rural Development as a national standard, numbered GB50016-2014. It was the first unified general Fire Code for Architectural Design in China. The new edition included 12 chapters, 3 appendices, and 425 articles.

On November 30, 2016, the newly formulated industry standard TB10063-2016 Fire Code for Railway Engineering was approved by the National Railway Administration; it came into force on March 1, 2017. The 2016 edition was developed based on the 2012 edition according to the characteristics and practices of railway engineering, the latest performance-based fire safety engineering developments in railways, and the relevant recently promulgated national laws and regulations. It is the new technical standard on fire safety engineering in railways.

On May 27, 2017, the GB 51236-2017 Fire Code for Civil Airport Terminal Design was approved by the Ministry of Housing and Urban-Rural Development as a national standard; it came into force on January 1, 2018.

On March 30, 2018, the partially revised national standard GB50016-2014 Fire Code for Architectural Design (2018 Edition) was approved by the Ministry of Housing and Urban-Rural Development; it took effect on October 1, 2018. This edition adds technical fire safety requirements for elderly care facilities.

On May 14, 2018, the national standard GB 51298-2018 Fire Code for Metro Design was approved by the Ministry of Housing and Urban-Rural Development; it came into force on December 1, 2018.

3.2   Comparison Between US and Chinese Codes (Table 3.1)

Table 3.1 Comparison between US and Chinese codes of fire safety design for transportation hubs

Item		TB 10063	NFPA 130
Fire prevention zoning	Waiting halls and passenger service centers	Where waiting halls and passenger service centers meet certain conditions, the maximum allowable floor area of their fire prevention zones shall not exceed 10,000 m²	Fire prevention zones have no area limitation and shall be compartmented according to their functions
		Where railway passenger stations are jointly constructed with other buildings, separate fire prevention zones shall be designed	
		Public passenger areas of medium-sized and larger railway passenger stations and their centralized office areas and equipment areas shall be divided into separate fire prevention zones; the fire prevention zones have no area limitation and shall be compartmented according to their functions	
Fire compartmentation	Transformer substations	6.2.1 The following building structures shall be separated from other parts by partitions with a fire resistance rating of at least 2.00 h, and floors with a fire resistance rating of at least 1.50 h. Doors and windows connecting these structures with other parts shall be Class B fireproof doors and windows:	Transformer substations shall be separated from the surrounding spaces by 3.0 h fireresistive construction
	Electrical control rooms, auxiliary electrical rooms, and related battery rooms	1. Communication rooms at railway communication centers; communication rooms at dispatching centers (stations); communication rooms at stations; communication rooms at railway sections (communication machinery rooms at base stations, relay stations, and traction power and power stations (houses))	Electrical control rooms, auxiliary electrical rooms, and related battery rooms shall be separated from the surrounding spaces by 2.0 h fireresistive construction

(continued)

Table 3.1 (continued)

Item		TB 10063	NFPA 130
	Garbage rooms	2. Equipment rooms at dispatching centers (stations) and signal machinery rooms (including signal equipment rooms, relay rooms, power supply rooms, and lightning protection rooms) and operation rooms at stations, bullet train sections (stations), and railway sections	Garbage rooms shall be separated from the surrounding spaces by 1.0 h fireresistive construction
	Vehicle control rooms and related battery rooms		Vehicle control rooms and related battery rooms shall be separated from the surrounding spaces by 2.0 h fireresistive construction
		3. Information equipment rooms and fire control rooms	
	Public areas, nonpublic use areas	4. Vehicle safety warning system rooms	Public areas shall be separated from the nonpublic use areas by 2.0 h fireresistive construction
		5. Central computer rooms monitoring natural disasters and foreign matter intrusion	
	Public application areas, non-transportation structures	6.2.2 Main control rooms, distribution device rooms, compensation device rooms, and transformer rooms at traction substations, section substations, autotransformer substations, and switching stations as well as control rooms at 10 kV and above transformer and distribution substations shall be separated from other parts by partitions with a fire resistance rating of at least 2.00 h and floors with a fire resistance rating of at least 1.50 h Where main control rooms, distribution device rooms, compensation device rooms, and transformer rooms at traction substations, section substations, autotransformer substations, and switching stations as well as control rooms at 10 kV and above transformer and distribution substations are jointly constructed with other buildings, and their internal doors and windows shall be Class A fireproof doors and windows	Public application areas shall be separated from the non-transportation structures by 3.0 h fireresistive construction

(continued)

Table 3.1 (continued)

Item		TB 10063	NFPA 130
		6.2.3 Cable shafts of communication rooms, signal machinery rooms, information equipment rooms, dispatching centers (stations), vehicle safety warning system rooms and transformer and distribution substations, traction substations, section substations, autotransformer substations, and switching stations shall be equipped with an enclosure structure with a fire resistance rating of at least 1.00 h and Class B fireproof access doors	
Evacuation	Occupant load	——	The occupant load for a station shall be based on the vehicle load of vehicles simultaneously entering the station on all tracks (a peak period lasts for 15 min) plus the load of passengers
	Evacuation distance	The Fire Code for Architectural Design stipulates that "the straight distance from any point in the room to the nearest evacuation door or safety exit shall not exceed 30 m; where the evacuation door cannot be directly connected to the outdoor surface or the evacuation staircase, an evacuation corridor with a maximum length of 10 m that leads to the nearest safe exit shall be provided. Where an automatic fire sprinkler system is installed in this place, the safe evacuation distance from any point in the room to the nearest safe exit can be increased by 25%"	The maximum travel distance from any point on the platform to the exit shall not exceed 100 m (325 ft)

(continued)

Table 3.1 (continued)

Item		TB 10063	NFPA 130
	Evacuation width	Evacuation doors in densely populated public places and auditoriums shall have a minimum clear width of 1.40 m and have no thresholds or steps within 1.40 m inside and outside the doors. Outdoor evacuation corridors in densely populated public places shall have a minimum clear width of 3.00 m and directly lead to spacious areas	A minimum clear width of 1120 mm (44 in.) shall be provided along all platforms, corridors, and ramps serving as means of egress
			5.3.4.2 In computing the means of egress capacity available on platforms, corridors, and ramps, 300 mm (12 in.) shall be deducted at each sidewall, and 450 mm (18 in.) shall be deducted at platform edges that are open to the trainway
			5.3.4.3 The maximum means of egress capacity of platforms, corridors, and ramps shall be computed at 0.0819 p/mm-min (2.08 p/in.-min)
			5.3.4.4 The maximum means of egress travel speed along platforms, corridors, and ramps shall be computed at 37.7 m/min (124 ft./min)
			5.3.4.5 The means of egress travel speed for concourses and other areas where a lesser pedestrian density is anticipated shall be computed at 61.0 m/min (200 ft./min)
Fire safety facilities	Automatic fire alarm system	9.1.1 An automatic fire alarm system shall be provided at the following locations:	

(continued)

Table 3.1 (continued)

Item		TB 10063	NFPA 130
		1. Places with an automatic gas extinguishing system and automatic fire sprinkler system (excluding tunnel equipment rooms)	
		2. Logistics center warehouses, luggage express transportation bases, station cargo warehouses, and luggage and package warehouses with a floor area of more than 1000 m^2	
		3. Main equipment rooms at traction substations, section substations, autotransformer substations, and switching stations, including communication machinery rooms, distribution device rooms, combustible medium compensation device rooms, control rooms, oil-immersed transformer rooms, and cable mezzanines and shafts	
		4. Centralized storage spaces for passenger trains at bullet train sections (stations), passenger train preparation, stations and passenger train maintenance stations	
		5. Integrated computer rooms, bill warehouses, and power distribution rooms at mega and large passenger stations and frontier (port) stations, joint inspection rooms, and fire-prone areas at frontier (port) stations	
		6. Locations that integrate an automatic fire alarm system with a mechanical smoke control system, a rain sprinkler or automatic pre-action sprinkler system, a fire plug system, and an automatic water injection system	

(continued)

Table 3.1   (continued)

Item		TB 10063	NFPA 130
	Indoor fire water supply	7.2.1 Indoor water fire supply shall be provided in the following buildings and the workshop hazard A, B, and C (defined by GB 50016 3.1.1) and warehouses with an area of more than 300 m² as specified in Appendix A of this Code:	
		1. Repair garages for diesel locomotives and repair and parking garages for large track maintenance machines	
		2. Traffic, locomotive, vehicle, office, electrical, and living buildings in the railway station area that provide railway transportation and production services and have a volume greater than or equal to 10,000 m³ or a height greater than 15 m	
		7.2.2 Indoor fire water supply shall not be required in the following buildings or locations, but other fire protection measures shall be taken:	
		1. For a station with no fire water supply, its fire danger rating level should be raised to the next higher level, and appropriate fire extinguishers shall be provided	
		2. For section stations, autotransformer substations, switching stations, relay stations, base stations, and other small signal, communication, and information equipment rooms without production and living water supply facilities, their fire danger rating level should be raised to the next higher level, and appropriate fire extinguishers shall be provided	

(continued)

Table 3.1 (continued)

Item		TB 10063	NFPA 130
		3. A traction substation without fire water supply shall be equipped with two sets of mobile high-pressure water mist extinguishing devices	
		4. Railway garages and internal combustion forklift garages with six parking spaces or less shall be equipped with four 35 kg cart-type ABC dry powder extinguishers	
		7.2.3 Indoor fire water supply for underground stations shall comply with the GB 50157 Fire Code for Metro Design	
		7.2.4 Passenger service centers, ticket offices, and waiting halls (rooms) at passenger stations shall be provided with fire hydrant cabinets that contain fire hose reels	
	Smoke control system	The Fire Code for Architectural Design stipulates: 8.5.1 Smoke control facilities shall be installed at the following locations or parts of a building: 1. Smoke-proof staircases and their front chambers	7.1.2.1 For length determination, all contiguous enclosed trainway and underground system station segments between portals shall be included
		2. Front chambers or shared front chambers of fire lift 3. Front chambers of corridors of refuge and areas (rooms) of refuge 8.5.3 Smoke control facilities shall be installed at the following locations or parts of a civil building:	7.1.2.2 A mechanical emergency ventilation system shall be provided in the following locations: (1) In an enclosed station

(continued)

Table 3.1 (continued)

Item		TB 10063	NFPA 130
		1. Entertainment venues with a floor area of more than 100 m² located on the first, second, and third floors; entertainment venues located on the fourth floor or aboveground, underground or semiunderground 2. Atriums 3. Aboveground rooms with a floor area of more than 100 m² where people often stay in a public building 4. Aboveground rooms with a floor area of more than 300 m² that store a relatively large quantity of combustibles in a public building 5. Evacuation corridors with more than 20 m in length in a building 8.5.4 Smoke control facilities shall be provided in underground or semiunderground buildings (rooms) and windowless rooms in aboveground buildings when the total floor area of the building is more than 200 m² or the floor area of the room is more than 50 m² and the building or room often has people staying there or stores a relatively large quantity of combustibles	(2) In an underground or enclosed trainway that is greater in length than 1000 ft. (305 m) 7.1.2.3 A mechanical emergency ventilation system shall not be required in the following locations: (1) In an open system station (2) Where the length of an underground trainway is less than or equal to 200 ft. (61 m) 7.1.2.4 Where supported by engineering analysis, a nonmechanical emergency ventilation system shall be permitted to be provided in lieu of a mechanical emergency ventilation system in the following locations:
			(1) Where the length of the underground or enclosed trainway is less than or equal to 1000 ft. (305 m) and greater than 200 ft. (61 m) (2) In an enclosed station where engineering analysis indicates that a nonmechanical emergency ventilation system supports the tenability criteria of the project

3.3 Performance-Based Fire Safety Design for Large Transportation Hubs

3.3.1 Proposal of Performance-Based Fire Safety Design

1. Fire Safety Design for Buildings

Fire safety design for buildings is an engineering act: it's the process of using certain methods to determine fire protection measures according to the fire protection requirements for the buildings. Building fire safety measures include reduced fire hazards, rapid response, timely and efficient fire extinguishing, and building structures meeting the fire resistance requirements, which allow quick and safe evacuation of people in an effective manner.

Passive and active fire protections are the main fire protection measures for buildings. Passive fire protection strategies include fire resistance and fire insulation technology, while active fire protection strategies include smoke control measures, safe evacuation, and provision of fire safety facilities (Fig. 3.1).

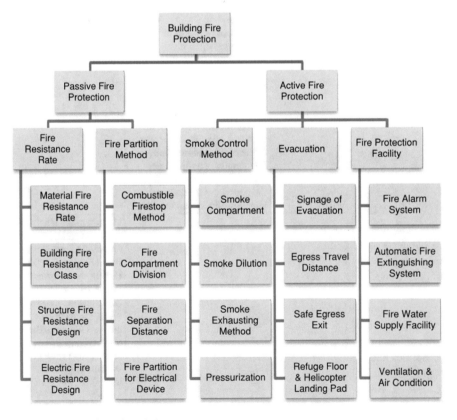

Fig. 3.1 Building fire safety design system

The traditional fire safety design for a building is based on the design parameters, standard requirements, and fire safety facility provision standards calculated in accordance with the type, height, and floor number of the building as stipulated in the relevant fire code for architectural design.

In order to ensure that buildings in the whole country or some large areas reach the predetermined safety level, after years of study and practices, the Ministry of Housing and Urban-Rural Development of China has promulgated the Fire Code for Architectural Design. The code, which applies to both civil and industrial buildings, specifies the basic safety technology requirements for new construction and reconstruction projects and the mandatory safety requirements for fire safety design. Most of the provisions are mandatory and must be observed by the designers.

Rule-based regulations have a long history in China and have certain advantages: they are easy to apply due to easily executable parameter provisions; they can be easily understood by well-trained construction inspectors and skilled technicians.

However, as society advances, rule-based regulations have reflected many problems:

(1) Not conducive to innovative design

As rule-based regulations lack flexibility, new design concepts and new materials cannot be used due to conflict. The innovation of public environment and safety protection in architectural design needs to be driven by performance-based regulations.

(2) Unreasonable construction cost control

For special buildings, such as large shopping malls, gymnasiums, and high-tech semiconductor plants, rule-based regulations cause a large amount of unnecessary construction costs.

(3) Not conducive to international trade in building materials

As rule-based regulations restrict the types and specifications of materials and the use of certain types of regional materials, they contribute negatively to the use of foreign materials. In this regard, the International Trade Organization (ITO) writes: "Technical regulations should be based on product performance rather than design or descriptive characteristic."

(4) Design standards unable to keep pace with the times

Modern architectural trends develop constantly, while design codes remain unchanged. With the emergence of intelligent, multipurpose, mega high-rise, and underground buildings, regulations can hardly apply to these special types of buildings. In China, fire regulations will only be revised after the occurrence of major fire accidents. In addition, as the Ministry of Public Security and fire authorities, which are responsible for developing fire regulations, and architects, who are responsible for architectural design, vary widely in their expectations of building fire safety principles and standards and desired outcomes, sometimes the latter exploits regulation loopholes to achieve the expected architectural effect.

2. Proposal of performance-based fire safety design

The concept of performance was put forward as early as in the first human building code at the time of King Hammurabi of Babylon B.C., in which Article 229 stipulated that "The builder must ensure the structural safety performance of a building." A 1925 study by the US National Bureau of Standards, titled Recommended Practice for Arrangement of Building Codes, states: "Requirements should be stated in terms of performance, based upon test result for service conditions, rather than in dimensions, detailed methods, or specific materials."

Performance-based fire safety design is a fire safety plan that designers create based on the building's space characteristics by using fire safety methods to meet the goal of fire safety. Performance-based design is characterized by goal and function orientations, consideration of different fire risk factors, flexibility to choose to meet the code or the building's actual situation, and consideration of cost-effectiveness.

Performance-based design has the following advantages:

(1) The design plan is more reasonable as a variety of analysis software is used to improve its accuracy and excellence; it is more relevant to different building forms; and it is more flexible as it is developed based on actual needs.
(2) The design plan is more economical and energy-efficient and requires lower construction costs.
(3) It contributes positively to the application of new materials and technologies.
(4) It takes the building fire safety system as a whole into consideration, coordinating various fire safety systems.

3. Comparison between performance-based design and rule-based regulations

Performance-based and rule-based fire safety designs share the same goal – achieving building fire safety – but also show differences and complementarity (Table 3.2).

Table 3.2 Comparison between performance-based and rule-based fire safety designs

Item	Traditional fire safety design	Performance-based fire safety design
Economy	Economical	Expensive
Practicality	Simple	Complex
Design and assessment cycle	Short	Long
Design and assessment cost	Low	High
Design plan	Simple	Diversified
To new technologies, processes, and products	Slow response, application limitations	Emphasis on individuality, easy application
Areas of advantage	Traditional buildings	New buildings
Building quantity to which the design applies	Large quantity	Small quantity
Requirements for personnel quality	Low	High
Present application status	Main widely used method	New less-used method
Legal basis	Available	Currently unavailable
Standard system	Complete	Yet to be developed

3.3.2 Process and Objectives of Performance-Based Fire Safety Design for Large Transportation Hubs

1. Procedures for evaluating performance-based design for large transportation hubs

Due to different viewpoints and analysis habits, there is currently no unified process for performance-based fire safety design. For example, the SFPE (US Society of Fire Protection Engineers) Engineering Guide to Performance-Based Fire Protection Analysis and Design of Buildings provides a nine-step performance-based design process, while the BS (British Standards) DD240 Fire Safety Engineering in Buildings presents a four-step design approach. Despite different design processes, these design methods are based on performance-based quantitative analysis and have the same basic contents. In China, the performance-based fire safety design for large transportation hubs may be divided into 11 steps according to the main contents of performance-based fire safety design and the characteristics of these buildings themselves (Fig. 3.2).

2. Large transportation hub design process (Fig. 3.3)

3. Objectives of performance-based fire safety design for large transportation hubs

Performance-based fire safety design objectives are divided into fire safety objectives and fire safety design objectives.

Performance-based safety objectives focus on the overall ultimate effect of a safety system, while fire safety objectives use analysis and design as a starting point. Performance-based objectives consist of three levels. For a special type of large public buildings such as large transportation hubs, the primary goal of fire protection is to control fire quickly and effectively and ensure the evacuation of trapped people in an effective manner (Table 3.3).

3.3.3 Performance-Based Fire Safety Design Approach for Large Transportation Hubs

1. Performance-based quantitative analysis

Performance-based quantitative analysis involves setting fire safety objectives of a building, establishing typical building fire scenarios, simulating smoke control and safe evacuation in typical fire scenarios through building fire safety assessment, obtaining quantitative analysis results, and verifying whether the preliminary performance-based fire safety design plan can achieve the set objectives and ensure safe evacuation and the structural safety of the building.

In terms of safe evacuation, the performance-based approach involves analyzing the typical fire scenarios and evacuation scenarios through computer software simulation and obtaining and comparing the available safe egress time (ASET) and the required safe egress time (RSET). When ASET >1.2 RSET, it means that under the

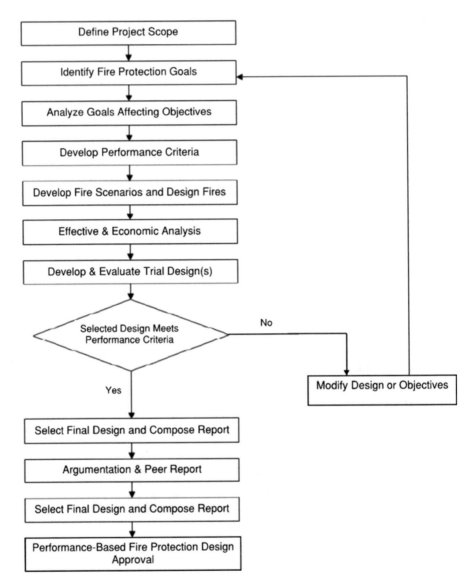

Fig. 3.2 Performance-based design process of large transportation hubs

given fire scenario, the preliminary fire safety plan can achieve the goal of safe evacuation within a time period that ensures life safety. Otherwise, it means that people cannot be evacuated to a safe area within the effective time period and that the preliminary plan is not feasible and needs further modification (Figs. 3.4 and 3.5).

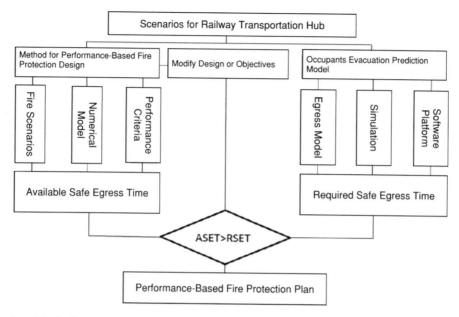

Fig. 3.3 Performance-based fire safety design flowchart

Table 3.3 Performance-based fire safety design objectives

Performance-based fire safety design objectives		Contents
Fire safety objectives	Social assessment objectives	Protecting life and property safety, functions of buildings, or continuity of services, preventing the environment from harmful effects of fire, etc.
	Functional assessment objectives	To reach the goal of life safety, the buildings and their systems must be functional enough to ensure safe evacuation of people in case of fire
	Performance objectives	Representing the performance requirements for buildings and their systems. In order to achieve the overall fire safety objectives and functional objectives, building materials and elements, system components, and building methods must meet certain performance requirements. Performance objectives will present requirements for fire separation, fire detection and alarm systems, smoke control systems, and even automatic fire sprinkler systems of buildings
Fire safety design objectives		Avoiding fire
		Fire control
		Avoiding losses
		Completing evacuation

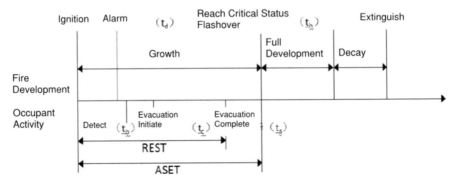

Fig. 3.4 Time line for fire growth and evacuation

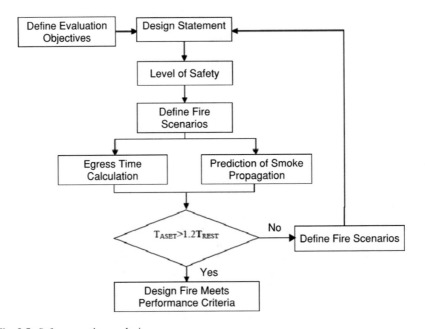

Fig. 3.5 Safe evacuation analysis

2. Critical state

The critical state in a fire accident refers to the state in which the fire environment can cause serious injury to indoor personnel. The factors to consider when determining the danger a fire poses to people are the thermal radiation of the flame and smoke layer, the concentration of toxic gases in smoke, and the high temperature and visibility of smoke. According to the requirements of relevant safety standards, the criteria for personnel safety evaluation during evacuation are provided below (Table 3.4):

Table 3.4 Criteria for personnel safety evaluation

Item	Criteria
Thermal radiation (hot smoke layer)	At an altitude of more than 2.0 m, the hot smoke layer concentration or temperature shall be less than 2.5 kW/m² or less than 180 °C for ordinary people and shall be less than 10 kW/m² or less than 375 °C for fire-fighters
Convective heat flow	At an altitude of less than 2.0 m, the ambient temperature shall not exceed 60 °C for ordinary people and shall not exceed 260 °C for fire-fighters
Visibility	At an altitude of less than 2.0 m, visibility shall be at least 10 m
CO concentration	At an altitude of less than 2.0 m, CO concentration shall not exceed 450 ppm for ordinary people
CO_2 concentration	At an altitude of less than 2.0 m, CO_2 concentration shall not exceed 1% for ordinary people

3.3.4 Important Concepts in Performance-Based Design

A large transportation hub integrates various means of transport and has complex functions and internationally cutting-edge designs. Due to its unique functional requirements and architectural characteristics, conventional fire protection codes cannot cover its fire safety design requirements. Scientific introduction of performance-based fire safety design and prior case analysis can better achieve the principles of safety, reliability, advanced technology, and economic viability. This section will focus on some important concepts in performance-based fire safety design, which have been used many times in international and domestic large-scale projects and play a guiding role in fire safety design for large transportation hubs.

1. Fire barrier belt

A fire barrier belt is an area with a certain width between different functional halls where combustibles are strictly forbidden. The effective width of the fire barrier belt is calculated using a radiation model to ensure that the flame on one side does not spread to the other side in case of fire. The concept is generally used in high and large spaces, where physical measures such as fire walls can be avoided by setting up a fire barrier belt as the former seriously affects the traffic function and the architectural effect.

A fire barrier belt shall have smoke walls (a mechanical smoke control system) preventing the spread of smoke and shall be equipped with a separate fire prevention system, such as smoke control and sprinkler systems or lighting and ventilation shafts. In order to avoid the spread of fire, the suspension ceiling in the fire barrier belt shall be provided with fire partitions, and a pipe crossing a partition shall have a fire valve.

In project design, a fire barrier belt shall be provided with a wet alarm valve, a separate sprinkler system, and a smoke control system (or a ventilation shaft), which shall be located higher than the smoke walls. Therefore, even if some smoke spreads across the smoke wall from one side of the fire barrier belt, it can be vented through the smoke control system and has little possibility of spreading beyond the other side of the fire barrier belt.

Fire zones separated by "fire barrier belts" are also called logical fire zones. The concept is widely used in forests and large factories and also in public transportation hubs such as large airports and railway stations (Fig. 3.6).

2. Fire cabin

The so-called fire cabin is composed of a solid roof, covering areas with high fire loads, such as retail areas, ticket offices, and business office areas. Automatic alarm and sprinkler systems and mechanical smoke devices are installed under the roof. In this way, fire can be quickly suppressed, and smoke can be prevented from spreading to large spaces (Fig. 3.7).

By introducing the concept of performance-based "fire cabin" to large space fire safety design, mechanical smoke control, automatic alarm and sprinkler systems, and fire compartmentation can be integrated into fire-prone areas, minimizing the occurrence of incidents endangering life, property, and operations safety. In this way, it's unnecessary to physically separate a large space to limit the spread of fire and smoke, thus ensuring the free flow of personnel and the continuity of operations. The concept of "fire cabin" has been tested and applied successfully in many large international airports and stations around the world.

Fire cabins come in two types: open fire cabins and enclosed fire cabins. An open fire cabin shall have a smoke canopy at the top, but is not required to have fire walls on four sides. When used in retail areas, an open fire cabin often has open sides, except partitions between stores. An enclosed fire cabin may have four closed sides or one open side that automatically closes when detecting a fire. An enclosed fire cabin may be used in a retail area adjacent to a key area where the occurrence of any fire will seriously impact the continuity of operations of the transportation hub (Figs. 3.8 and 3.9).

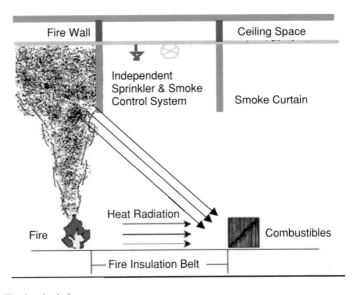

Fig. 3.6 Fire barrier belt

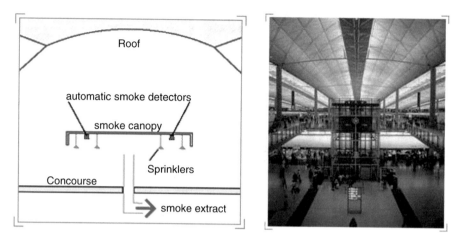

Fig. 3.7   Practical application of fire cabin at Hong Kong International Airport

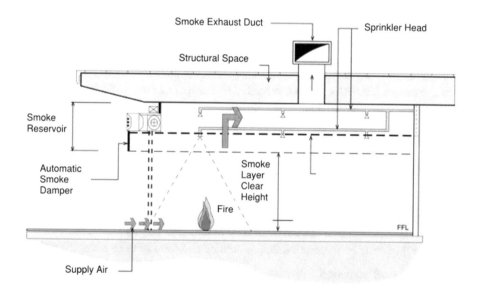

Fig. 3.8   Open fire cabin

3. Fuel island

   For a large transportation hub that may have several mobile or fixed kiosks and business office areas, these kiosks and office areas need to be protected by "fuel islands" when fire cabins are unavailable. That is to say, these kiosks and business areas should be regarded as isolated fuel areas.

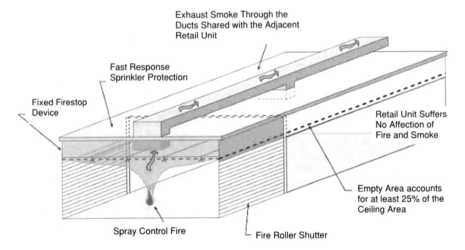

Fig. 3.9 Enclosed fire cabin

When the fire load directly facing a large space burns, flue gas and nearly two-thirds of the heat will spread toward the high roof through convection. Therefore, the spread of a fire between combustibles mainly depends on thermal radiation of the fire source. The effective radius ignited by thermal radiation of a fire source in a large space is usually calculated by the following formula:

$$R = \left(\frac{Q}{12\pi q''} \right)^{1/2}$$

where:

Q is the heat release rate of the fire source, kW.
R is the effective ignition radius calculated from the fire source center, m.
q'' is the minimum heat radiation flux ignited by radiation of the igniting material.

The thermal radiation intensity of a fire source decreases sharply with the increase of distance and the decrease of the heat release rate of the fire source. Therefore, once an adequate distance is guaranteed between the fire load of combustibles and the combustibles themselves, fire spread will usually not occur even without the protection of a sprinkler system. In this way, combustibles form a fuel island and the traffic area between islands become a natural fire barrier belt.

It is necessary to control and protect the fire load of fuel islands. The introduction and storage of dangerous goods shall be prohibited in fuel islands. In particular, the introduction of a large amount of combustibles shall be avoided as much as possible to prevent a large fire and an excess width of the fuel island that affects architectural and functional design.

4. Fire unit

A large space often prohibits the provision of fire walls and curtains or other facilities in accordance with the fire zoning requirements of fire codes. But for some electromechanical and auxiliary rooms, the concept of fire unit can be used to protect and separate them from the whole large space.

A fire unit protects rooms with local maintenance structures within a large space using 2 h fireresistive partitions and Class B fireproof doors and windows on the partitions. A fire unit should cover an area of less than 2000 m². As a fire unit is often in a large space and mainly houses staff and other larger areas with high fire loads have special protection measures, the concept of fire unit provides flexibility for design and facilitates specific, well-arranged fire safety design for large spaces.

5. Egress passageway

Since some evacuation staircase cannot directly connect to outdoor and in order to provide safe environment for evacuation, "egress passageway" is set to guide people to the outdoor safe area. The design of it should meet the following requirements:

(1) The fire resistance limit of the firewall of the egress passageway should not be lower than 3.0 h, and the fire resistance limit of the floor slab should not be lower than 1.50 h.
(2) The distance between the door of any fireproof zone leading to the egress passageway and the nearest exit of the passageway to the ground shall not be greater than 60 m. The net width of the egress passageway shall not be less than the designed total net width for evacuation of the design evacuation of any fire zone leading to the egress passageway.
(3) The combustion performance of the interior decoration material of the egress passageway should be A-grade.
(4) Smoke-proof front room should be set up at the entrance of the egress passageway from the fireproof partition, the use area of the front room should not be less than 6.0 m², the door opening to the front room should adopt Class A (1.5 h) fireproof door, and the door of front room opening to the egress passageway should adopt Class B (1.0 h) fireproof door.
(5) A front room should be set up at connecting part of parking garage to the egress passageway, and the area should not be less than 6.0 m².
(6) The automatic fire alarm system, fire hydrant, fire emergency lighting, emergency broadcast, and fire line telephone should be set up in the egress passageway (Fig. 3.10).

3.3.5 Important Parameters of Performance-Based Design

1. Evaluation criteria for safe evacuation

An important purpose of safe personnel evacuation analysis is to determine whether the whole evacuation process is safe or not by calculating the available safe

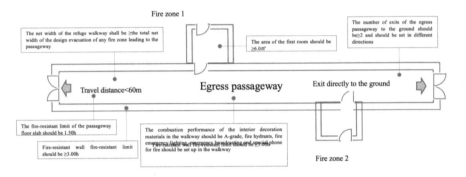

Fig. 3.10 Egress passageway

egress time (ASET) and the required safe egress time (RSET). In order to determine that a performance-based fire safety design can ensure safe evacuation, it is necessary to prove that the trapped people can reach a safe site before suffering from injury caused by fire.

The rules in the Australian BCA Code are currently internationally accepted and widely recognized standards for safe evacuation. This code requires the ASET to be 1.2 times greater than the RSET to guarantee safe evacuation. The RSET consists of detection time (T_{cue}), occupant response time (T_{reso}), and movement time (T_{trav}).

ASET (available safe egress time) = T_{cue} (detection time) + T_{reso} (occupant response time) + T_{trav} (movement time) (Fig. 3.11)

Detection time and occupant response time are determined by code requirements and data accumulation. Movement time is calculated by simulating fire evacuations from different locations including outbound, platform, and elevated floors with software like Pathfinder and buildingEXODUS.

The effective evacuation time after fire, or ASET, is determined through systematic simulation and analysis of smoke movement, taking into consideration various factors.

Smoke simulation and analysis is the method of using computational fluid dynamics (CFD) to study fire occurrence and spread and smoke movement in large transportation hubs under the given fire scenarios.

2. Detection time

Detection time is the sum of reaction time of fire detection systems and notification time after fire confirmation. The densely populated areas in railway stations and integrated large transportation hubs can be roughly classified into the following two types by detection time:

(1) Outbound floors, passenger service halls, elevated floors, and mezzanines

As no time is specified in relevant Chinese codes, the US NFPA 72 (National Fire Alarm and Signaling Code) is used here, which provides 3 min or 180 s detection time for the station outbound floors, passenger service halls of eastern and western station buildings, and elevated floors and mezzanines.

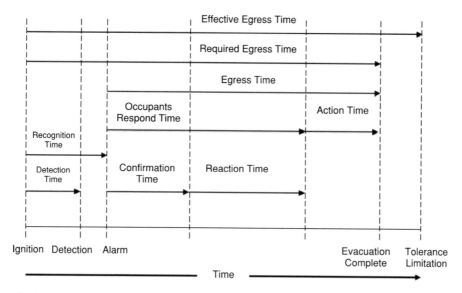

Fig. 3.11 Time distribution for safe evacuation in fire

(2) Platform floors

For a platform floor, a fire will mainly occur on a train at the platform. As the platform floor is open and doesn't obstruct sight, platform personnel can quickly learn about the occurrence of a fire by noticing the light and smoke. So, the detection time is about 60 s.

3. Occupant response time

The occupant response time is the time required for deciding the direction of evacuation after the internal station staff receive a fire alarm notice or make an evacuation decision. It consists of confirmation time and response time:

(1) Confirmation time

After receiving the fire signal from the fire alarm system or detecting obvious flame and smoke, the internal building staff realize, after some delay, that they must evacuate. The amount of time delayed is affected by many factors, such as the environment the staff are in, the activities underway, the situation of the staff themselves, and the type of signals issued by the fire alarm system. The following table shows the occupant response time to different alarm systems in different types of buildings (Table 3.5).

The alarm system types listed in the table are:

W1 – Live broadcasting from CCTV control rooms or real-time instructions from well-trained uniformed staff members, which can be seen or heard by all occupants on the scene
W2 – Nondirective voice information, such as prerecorded voice broadcasting system, or visual alarm signs and trained staff

Table 3.5 Occupant confirmation time to various alarm systems in different buildings

Building use and characteristics	Confirmation time (min)		
	Alarm system types		
	W1	W2	W3
Office buildings, commercial or industrial buildings, schools (residents are awake and familiar with the buildings, alarm systems, and evacuation measures)	<1	3	>4
Stores, exhibition halls, museums, recreation centers, and other public buildings (residents are awake, but unfamiliar with the buildings, alarm systems, and evacuation measures)	<2	3	>6
Hotels or boarding schools (residents may be asleep and are familiar with the buildings, alarm systems, and evacuation measures)	<2	4	>5
Hotels and apartments (residents may be asleep and are unfamiliar with the buildings, alarm systems, and evacuation measures)	<2	4	>6
Hospitals, sanatoriums, and other public institutions (a considerable number of people need help)	<3	5	>8

W3 – An alarm system using alarm bells or other similar alarm devices

As large transportation hubs belong to the second type of public buildings listed in the table above, their alarm systems are basically Type W1, supplemented by well-trained management personnel in each area, so the confirmation time should be far less than 2 min.

(2) Response time

The response time is the time required for making preparations for an evacuation after the staff have received the fire alarm signal and realized a fire has occurred. For railway stations and integrated large transportation hubs, as passengers can easily notice flame, smoke, and other signals due to an open space and will be guided by broadcasts and trained staff, the response time is dramatically shortened and can be calculated as about 2 min.

4. Number of people to be evacuated

(1) Number of people to be evacuated from the waiting area

According to the statistical data, passengers' stay time in the waiting hall is about 20–40 min. Using the instantaneous flow rate calculated from the peak-hour delivery volume and an amplification coefficient of 1.1–1.4, the following formula is obtained:

Number of passengers to be evacuated = number of passengers per peak hour × amplification coefficient × stay time (min) / 60

After using the above formula to get the number of passengers to be evacuated from the waiting area, it should be added to about 10% people seeing off passengers and 10% service staff. Then, this total number of people to be evacuated will be horizontally compared with the maximum number of railway passengers gathered; the bigger number will be the final number (Table 3.6).

Table 3.6 Passenger proportion (%) by type in different types of waiting areas (rooms)

Building size	Waiting area (room)				
	Ordinary	Soft seats or berths	VIPs	Soldiers (groups)	Barrier-free
Mega station	87.5	2.5	2.5	3.5	4.0
Large station	88.0	2.5	2.0	3.5	4.0
Medium-sized station	92.5	2.5	2.0	——	3.0
Small station	100	——	——	——	——

(2) Number of people to be evacuated from the platform floor

The number of passengers to be evacuated from the platform floor should be the peak number of passengers on the platform. If two trains enter the platform at the same time and each train has 16 carriages (the maximum personnel quota for each carriage is 110), then the maximum passenger number of the two trains is 3520. In addition, 5% people picking up and seeing off passengers and railway service staff should be considered.

(3) Number of people to be evacuated from the outbound area

Determine passengers' stay time in the outbound area according to the statistical data, calculate the instantaneous flow rate based on the peak-hour delivery volume, and consider an amplification coefficient of 1.1–1.4 as well as 10% people picking up passengers and 10% service staff. Then, the number of people to be evacuated from the outbound area can be obtained.

(4) Number of people to be evacuated from the commercial mezzanine

If commercial facilities are installed in the mezzanine, the number of people to be evacuated should be calculated according to the relevant regulations on commercial buildings.

5. Occupant composition by type

The internal occupants of large transportation hubs can be roughly classified into adult males, adult females, and others including the weak and elderly. The occupant compositions for different public buildings are provided below according to the recommended ratios (Table 3.7).

6. Occupant walking speed

The internal occupant density of large transportation hubs is generally between 0.54 and 3.8 persons/m², and the corresponding walking speed is provided below (Tables 3.8 and 3.9):

7. Effective width of safety exits

The latest research report shows that internal occupants would keep a certain safe distance from the building components they encountered in evacuation process. In highly populated railway stations and integrated large transportation hubs, not the evacuation width of each building component can be fully utilized in evacuation process, so it shall be reduced accordingly (Table 3.10).

Table 3.7 Occupant types and compositions

Location type	Adult males	Adult females	Children	Seniors
Stores, pedestrian streets	40%	40%	10%	10%
Staff of various functional rooms	60%	40%	0%	0%
Movie houses	40%	40%	10%	10%

Table 3.8 Evacuation speed determined by SFPE

Occupant density (person/ m²)	<0.54	0.54–1	1–2	2–3	3–3.8
Horizontal evacuation speed (m/s)	1.2	1.2–1.0	1.0–0.66	0.66–0.28	0.28–0
Downstairs speed (m/s)	0.86	0.86–0.73	0.73–0.47	0.47–0.20	0.20–0

Table 3.9 Evacuation speed and physical characteristics of occupants

Occupant type	Walking speed (m/s)		Stores, horizontal corridors, entrances and exits	Body size (shoulder width (m) × back thickness (m) × height (m))
	Ramps and stairwells			
	Upward	Downward		
Adult males	0.5	0.7	1.0	0.5 × 0.3 × 1.7
Adult females	0.43	0.6	0.85	0.45 × 0.28 × 1.6
Children	0.33	0.46	0.66	0.3 × 0.25 × 1.3
Seniors	0.3	0.42	0.59	0.5 × 0.25 × 1.6

Table 3.10 Reduction in the effective width of different passageways

Passageway type	Reduction in effective width (cm)
Stairs, walls	15
Handrails	9
Concert hall seats, stadium benches	0
Corridors, ramps	20
Broad corridors, pedestrian corridors	46
Gates, arches	15

8. Fire size

In designing a fire safety system, the design fire size Q (MW) is an important parameter in the design of a smoke control system as it largely determines the smoke generation rate and is crucial and necessary for determining the area of natural exhaust outlets and the mechanical smoke exhaust rate.

The design fire size Q partly reflects the fire risk. Generally speaking, locations with an automatic sprinkler system usually have a smaller fire size, which means smaller fire risk. For locations without an automatic sprinkler system, if a fire is not timely and effectively contained, the fire size will mainly depend on the type and amount of fire loads and ventilation conditions. These locations usually have a larger fire size, which means larger fire risk.

Therefore, the fire type is closely related to the design fire size. Fire types usually include:

- Fuel-controlled fires
- Fires controlled by ventilation conditions
- Fires controlled by automatic sprinkler systems

(1) Steady-state fires

A steady-state fire is a fire that releases constant and maximum heat immediately after it starts. The size of a fire is usually determined by the maximum heat release rate of combustibles in the fire or the heat release rate upon start-up of a sprinkler system (if any).

In dealing with actual fires, the assumption of steady-state fire is considered quite conservative as the design fire scenario with a constant heat release rate ignores the growth and decline stages of the fire process.

(2) Growing fires

In reality, a fire usually experiences a growth stage, rather than immediately reaching the steady state with the maximum heat release rate. Fire growth is recognized as a function of time. Growing fires are often referred to as t-squared fires. The following formula is provided in the UK CIBSE TM19:

$$Q = 1000 \left(\frac{t}{t_s} \right)^2$$

where,

Q is the heat release rate, kW.
t is time, s.
tg is the standard growth time constant, s.

The t-squared fire formula is widely accepted and used to estimate real fires and considered to be suitable for defining fire growth under various conditions.

There is still some redundancy and conservativeness when using the t-squared fire formula. When a fire occurs, it may first enter the incubation or initial stage. The following figure taken from the US NFPA 92B shows a typical developing t-squared fire. The incubation stage of a fire is very difficult to predict, ranging from seconds to hours. In this analysis, assuming that a fire grows immediately after it occurs by completely neglecting the incubation stage is a highly conservative approach because the fire will usually cause smoke during the incubation stage due to smoldering, which is sufficient to activate the detection system or to be detected by personnel.

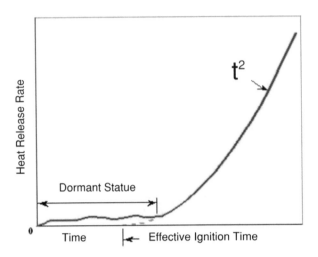

General true fire growth curve

T-squared design fires are divided by development speed into four widely recognized types: slow, medium, fast, and ultrafast. Table 3.11 shows the tg values of different types of fires, which indicate the time required for the fire to reach 1000 kW.

Compared with the steady-state fire model, the t-squared fire formula can reflect the actual situation of fires more accurately. Figure 3.12 shows the curves of several common t-squared fires.

Ultrafast fires are usually fires caused by plastic sheet products or fast-burning cushion furniture and gasoline dump fires (such as fossil oil and other flammable liquids). Fast fires are primarily store and hotel fires, while medium fires are mainly office and luggage fires.

(3) Sprinkler-controlled fires

An automatic sprinkler system is recognized as the most effective self-rescue facility in the world and the most widely used automatic fire extinguishing system. Domestic and foreign application practices prove that the system is safe, reliable, economical, and practical and has a high extinguishing rate.

The Australian Fire Engineering Guidelines recommend that the reliability coefficient of an automatic sprinkler system should be 0.95 before the "flashover" stage and 0.99 after the "flashover" stage in risk assessment. An automatic sprinkler system is regarded as a reliable fire extinguishing system.

When setting a sprinkler system, its start-up time in the designed fire scenario can be calculated using CFAST software. Generally, the fire heat release rate at the start-up time of the sprinkler system is regarded as the highest fire grade. In fact, an automatic sprinkler system has high extinguishing efficiency and can usually effectively control the fire once sprinkler heads start responding (Fig. 3.13).

The following assumptions are used in CFAST calculations:

- Start-up temperature of sprinkler heads: 68 °C
- Ambient temperature: 24 °C

Table 3.11 Types of fire growth

Type	tg (s)
Slow	600
Medium	300
Fast	150
Ultrafast	75

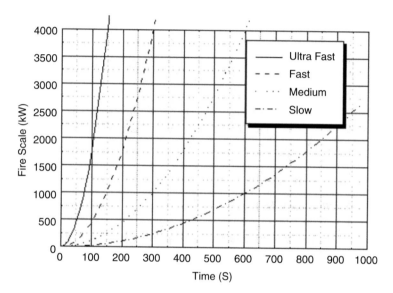

Fig. 3.12 T-squared fire growth curves

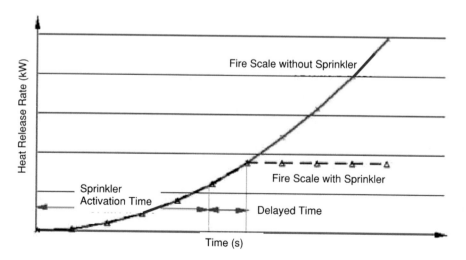

Fig. 3.13 Sprinkler-controlled fire size curve

- The longest distance between fire source and sprinkler heads: 2.5 m
- Quick response sprinkler head RTI = 50 m$^{1/2}$ s$^{1/2}$ (Quick response sprinkler heads are used in all areas of this project.)

The fire growth model and ceiling height, i.e., sprinkler head installation height, are determined according to different scenarios. At the same time, when determining the fire size in the design of a smoke control system, the factors causing delays in the start-up of the system must be considered, including:

- Delay in fire detection
- Delay in the alarm system
- Delayed closure of fire valves
- Delayed start-up of exhaust fans

In order to account for these system delays, a safety coefficient should be added to determine the design size of a sprinkler-controlled fire. In this design, a safety coefficient of 1.5 is used.

3.4 Chapter Summary

The chapter first describes the development process of US and Chinese fire safety design standards for large transportation hubs and compares the requirements of codes from the two countries to provide normative basis for fire safety design of large transportation hubs. Then, it provides the concept, process, objectives, and methods of performance-based fire safety design for large transportation hub.

Chapter 4
Fire Safety Strategies for Typical Space of Large Transportation Hubs

In recent years, construction of large transportation hubs in various countries around the world has been in full swing, among which China's large transportation hubs like those for high-speed trains, airports, and their composite buildings have been developing rapidly. The constant technological innovations of the Chinese high-speed train industry have led to the erection of one after another iconic high-speed train stations, further driving the development of the construction and design of large transportation hubs. Countries across the globe all have relevant standards for fire safety design, which is an important subproject of architecture design as it is directly related to people's life and property safety, such as the NFPA and China's Code for Fire Safety Design of Buildings. However, for large transportation hub buildings with great height, large space, and relatively complex functions, the fire safety design often cannot completely cover the existing standard codes. Therefore, such buildings normally adopt the performance-based design method to customize fire safety strategies. This book uses China's large transportation hubs as an example to systematically summarize the fire safety design of large transportation hub buildings.

4.1 Fire Compartmentation

4.1.1 Tradition Fire Compartmentation

The purpose of fire compartmentation in architecture design is to prevent fire from spreading. Fire compartmentation can be implemented for buildings according to the purpose and nature of rooms, and fire walls, fire doors, and fire-resistant shutters should be installed within the compartments.

In Tianjin West Railway Station, components like fire walls are used to separate the waiting area from functional rooms like ticket offices, shops, train inspection

© Springer Nature Switzerland AG 2021
F. Li, H. Li, *Fire Protection Engineering Applications for Large Transportation Systems in China*, https://doi.org/10.1007/978-3-030-58369-9_4

offices, and duty offices; walls with fireproof limit not less than 2.0 h are used to separate commercial and service rooms within the elevated waiting area; and the shops along the two sides of the entry level, commercial area along the two sides of the waiting hall, and shops and catering facilities within the mezzanine are separated from the waiting hall with fireproof glass that has fireproof limit of not less than 1.5 h (Fig. 4.1).

4.1.2 Dynamic Fire Compartmentation

In spaces with great height such as the entry and exit halls of large transportation hubs and waiting areas of airport terminals, it is difficult to install physical separation due to functional needs. As such spaces have relatively small fire load and great height, in the event of fire, heat dissipation is fast, and temperature rise is slow, so normally there is no flash-over, and the main hazard is the spread of smoke. Therefore, for fire compartmentation for such spaces, fire can be controlled by ensuring reasonable distance between combustible substances, setting fire isolation belts with a certain width, and enhancing smoke exhaust capacity and automatic fire extinguishing ability.

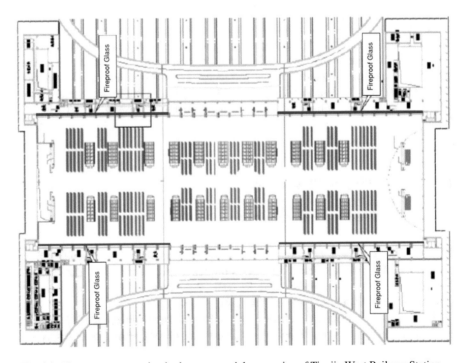

Fig. 4.1 Fire compartmentation in the commercial mezzanine of Tianjin West Railway Station

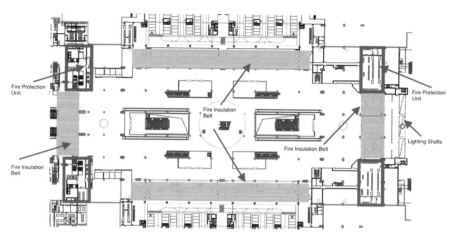

Fig. 4.2   Fire compartmentation on the −9.35 m level – Shanghai Hongqiao Eastern Transportation Square

A fire isolation belt is used between the metro station at Shanghai Hongqiao Eastern Transportation Square and the western airport terminal on the east. According to calculation, the width of the fire isolation belt should be 8 m, and the actual width is 18 m; the metro station is also separated with the maglev train station on the west by a fire isolation belt, the width of which should be 8 m according to calculation, and the actual width is 16 m; the metro station is separated from the garages on the south and the north with a ventilated courtyard and fire-resistant shutters, for which the width of the isolation belt should be 8 m according to calculation, and the actual width of the ventilation and lighting well is about 18 m; the entire ticket hall is a fireproof compartment, and equipment rooms in the area are set up as separate fire safety units according to standards (Fig. 4.2).

4.1.3   Super Large Fireproof Glass

The fire safety observation window for places like the station's computer room and control room needs to be transparent and have a wide vision, and there should not be blind spots due to lattice columns so that station staff can stay aware of what is going on outside the control room all the time; in terms of performance, the entire window should satisfy the requirements of fire safety integrity and 1.5-h heat insulation. Normally, a buffer area is set in front of the entrance to the central control room, which is equipped with security facilities. If the central control room and the emergency command room are next to each other, an observation window needs to be in place. The opening needs to have fire safety, which requires Class A (1.5 h) fireproof glass window.

Fig. 4.3 Super large Class A fireproof glass window of Xi'an Metro Line 4

Observation windows for metro projects including Xi'an Metro Line 4, Beijing Metro Line 8, and Wenzhou Metro Line S1 have passed inspection and put into operation, and the size of the largest fireproof glass is 3000 mm × 2000 mm (Fig. 4.3).

4.1.4 Traffic Tunnel Fire Compartmentation

Fire compartmentation in traffic tunnels is done according to functional zoning. Inside the tunnel, fire walls or fire retarding components with fireproof limit of not less than 3.00 h should be used to separate the tunnel's ancillary structures, service galleries as well as special fire shelters and evacuation passages, and independent fire safety refuge from the tunnel, forming a completely independent fire safety compartment.

The upline and downline highway transit level and the rail transit level of Shanghai Changjiang Tunnel both form independent fire safety compartments, which are separated by fire walls, and all the connection parts are equipped with Class A fireproof doors.

The circular tunnel section forms three independent fire safety compartments as shown below: the highway level, the lower-level reserved rail transit space (including safe passages), and the lower-level cable channel (Fig. 4.4).

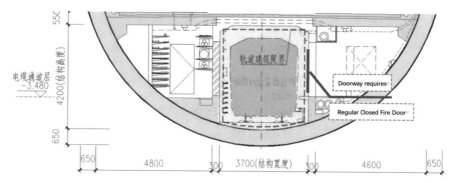

Fig. 4.4 Doorway closure

4.2   Smoke Control and Extraction Design

4.2.1   Smoke Control and Extraction Design for Large Transportation Hub Buildings

Tall and wide buildings have high ceilings and large spaces, which allow smoke to descend slowly and accumulate at the top of the building. In order to increase the time for evacuation in the event of fire, measures need to be strengthened to prevent, control, and mitigate the smoke descent.

Passenger entry/exit platform smoke control strategies at Foshan Western Railway Station ±0.000 m/5.200 m:

(1) Mechanical smoke extraction is not necessary for functional rooms smaller than 300 m².
(2) Set up mechanical smoke extraction systems in the central commercial area of the entry platform.
(3) Smoke extraction design for the passenger activity platform can adopt the following two options:

Option 1: The passenger entry platform, as a city traffic corridor, only has the people traffic and gathering function; its two ends on the south and the north are directly connected with the outdoor area, so natural ventilation and smoke exhaust can be achieved.

Option 2: Set up thrust mechanical smoke extraction systems on the north-south passenger platform on the 5.200 m exit level and above the ceiling of the exit hall, and mechanical smoke extraction systems should also be installed on the ±0.000 m level where the drop-off platform for taxis, buses, and private vehicles is located, and other spaces on the ±0.000 m platform can use the coverboard opening and the horizontal exit for natural ventilation and smoke exhaust.

The requirements for setting the thrust smoke extraction device include:

(1) The distance between the equipment and the structural wall or division wall should be no greater than 12 m.
(2) The horizontal distance between the equipment is no greater than 20 m (4 rows in lateral arrangement).
(3) The vertical distance between the equipment is no greater than 25 m (10 rows in longitudinal arrangement).
(4) The thrust of the equipment is no smaller than 100N.

Considering the operational needs for ventilation and cold smoke removal, it is recommended that thrust mechanical smoke extraction be adopted (Figs. 4.5, 4.6 and 4.7).

4.2.2 Smoke Control and Extraction Design for Transit Tunnels

Subway tunnels normally adopt vertical ventilation. When the train on fire cannot successfully enter the station and is forced to stop at the section, the fans at the two ends of the tunnel work together and form the longitudinal wind of no weaker than 2 m/s within the tunnel, causing the smoke to flow to one side and be exhausted to outdoors through the fan. The smoke control direction of longitudinal wind is not specified in both the domestic and foreign codes, which only stipulate that the direction of air supply should be the opposite of that of evacuation so that people can evacuate in the direction opposite to the direction the wind is blowing.

The normal direction of longitudinal ventilation inside the tunnel is either the same with or the opposite of the direction the train is going according to the position where the train catches fire. Salient piston wind is generated when the train is running in the tunnel, which controls the migration of the smoke when the train is moving and for a while after the train stops; thus, the smoke does not stay around the ignition point, making it very difficult to judge the ignition position of the train. [12]

An optimal smoke control plan is proposed for Shanghai Chongming Tunnel based on the smoke migration pattern in the event of train fire in the subway tunnel, that is, sections adopt the single longitudinal smoke control direction consistent with the direction the training is running. The pattern of smoke migration is as follows:

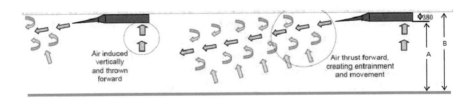

Fig. 4.5 Schematic diagram of working principle of thrust smoke extraction

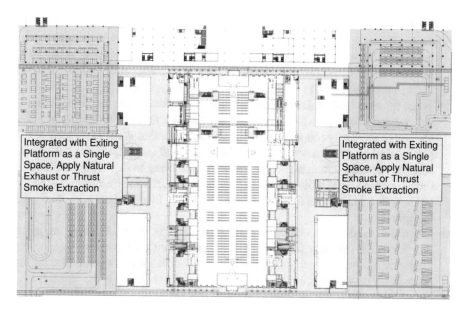

Fig. 4.6   Smoke extraction design on the passenger entry platform

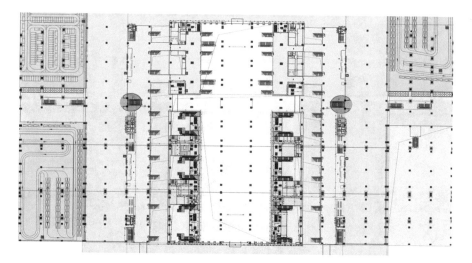

Fig. 4.7   Thrust smoke extraction design on the passenger exit platform

(1) The phase of natural smoke migration (no intervention of mechanical wind)
     The natural smoke migration process can be divided into three phases according
to the train speed and wind speed:

① When the train speed is greater than or equal to the wind speed, the direction of
   airflow velocity is from the headstock to the tailstock, and the control of the

piston wind over the smoke is far greater than that of the thermal pressure. When the ignition point is in the bottom outside of the train, the smoke "attaches" to the bottom of the train under the action of the piston wind and flows to the tailstock, where it follows the train and moves forward, forming a smoke column close to 100 m long following the tailstock, as shown in Figure (a).

② When the train brakes till the speed is 0, the piston wind in the tunnel slows down but does not stop immediately; instead, it continues to maintain a certain velocity, which gradually reduces under the action of friction drag. The speed of the piston wind is greater than the speed of the train, and the air flows from the headstock to the tailstock; meanwhile, hot smoke attaching to the bottom diffuses to the top of the tunnel while continuing to move toward the direction of the headstock, going against the direction of the piston wind for the first time, as shown in Figure (b).

③ When the velocity of the piston wind further reduces to close to zero, meaning 2–3 min after the train stops, the longitudinal wind's influence on the smoke gradually disappears, and when the piston wind weakens to the extent that it cannot restrain the impact of thermal pressure (the wind velocity in the tunnel is less than 1 m/s, equaling to annular wind speed of 2 m/s), the smoke newly generated from the fire source starts to diffuse to the interval tunnel on the two sides of the ignition point, the smoke in front of the headstock continues to move forward slowly, and the train's tailstock is gradually gulfed by the smoke, as shown in Figure (c).

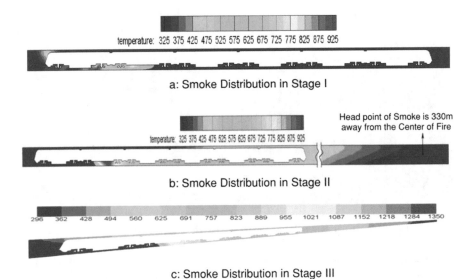

temperature: 325 375 425 475 525 575 625 675 725 775 825 875 925

a: Smoke Distribution in Stage I

temperature: 325 375 425 475 525 575 625 675 725 775 825 875 925

Head point of Smoke is 330m away from the Center of Fire

b: Smoke Distribution in Stage II

296 362 428 494 560 625 691 757 823 889 955 1021 1087 1152 1218 1284 1350

c: Smoke Distribution in Stage III

(2) The phase of mechanical ventilation intervention

When starting mechanical ventilation intervention, it has a significant impact on the shape of the smoke. The shape of smoke distribution is closely related to the timing and wind direction of mechanical ventilation intervention. Under normal circumstances, mechanical ventilation intervention should happen in Phase ②.

If the direction of the wind is the same with that of the train (hereinafter referred to as "forward ventilation"), the smoke shape distribution is similar to that of Phase ②, the smoke continues to move forward, and the high-temperature smoke generated by the fire source continues to add to the downstream smoke. If the direction of the wind runs against that of the train (hereinafter referred to as "reversal ventilation"), the direction of the smoke goes against that of the train for the second time, and the smoke moves toward the tailstock; the smoke column originally in front of the headstock flows back and passes the train for the second time and is ignited by the fire source, mixes with the smoke newly generated by the fire source, and moves toward the tailstock, as shown in Fig. 4.8. Whether the smoke will go against the direction of the train for the second time depends on the mechanical ventilation approach adopted.

(3) Smoke control strategy and application

When a running train catches a fire, the timing of mechanical ventilation should be determined according to the actual situation to make sure the ventilation direction follows that of the smoke and the impact is minimized (Table 4.1).

4.3   Application of the Cold Smoke Removal Function of Central Air Conditioning for Large Space

The high fire load areas of large space buildings are mostly protected by automatic fire extinguishing system. In the passage area with high ceiling and low fire load, even in the event of a fire, smoke will first gather under the tall and large ceiling, which ensures that people have sufficient time to evacuate, and the smoke will not be reduced to a height threatening people's safety in a short period of time. Therefore, the public flow area with low fire load in large space is generally not equipped with mechanical smoke exhausting system, but adopts the air conditioning system to remove the smoke after the disaster to restore normal operation on the site. The smoke control system is also known as the "cold smoke removal system" in that the cooling smoke is excluded after the fire. [10]

The smoke removal system should be activated by building HVAC managers or fire monitors. The smoke vent is usually the return air vent of the air conditioning system. When it comes to whether to set up mechanical smoke removal and the application of cold smoke removal system in large space, it is necessary to conduct

Head point of Smoke is 364m away from the Center of Fire

temperature: 325 375 425 475 525 575 625 675 725 775 825 875 925

Fig. 4.8   Smoke distribution after reversal ventilation

Table 4.1 Ventilation strategy

Serial no.	Phase of fire	Characteristics	Intervention measures	Remarks
1	Phase 1	The train speed is greater than the wind speed, and the smoke flows to the tailstock	Reversal ventilation	1. Valid ventilation is formed before the train reduces speed and stops 2. Difficult to implement, possible to implement with long sections, and almost impossible to implement in the subway
2	Phase 2	The train speed is smaller than the wind speed, and the smoke flows to the headstock	Forward ventilation	10–20 s after the train stops, open the doors at the tailstock for passengers to evacuate

FDS simulation experiments in advance, namely, to ensure that in the event of fire, even if the smoke is not removed, people can be safely evacuated within a certain period of time (at least 15 min). Since space has strong capacity to store smoke, only the post-disaster cold smoke removal is feasible.

The main building and promenade of T2 Terminal 13.60 m floor space adopt cold smoke removal system; ticket hall and international departure terminal have large space and strong ability to store smoke and heat; therefore, the concept of cold smoke removal is adopted, and the air conditioning system is combined with natural vents to exhaust smoke after the disaster. The smoke vent is the return air vent of the air conditioning system and is set at a height of 2–3 m. Take the ticket hall as an example: with a volume of about 1.2 million m³ and in order to resume the operation of the airport as soon as possible, at least one large space ventilation should be performed within 2 h, and wind volume for cold smoke removal should not be less than 600,000 m³/h.

4.4   Fire Safety for Steel Structure

Construction components of the space of large transportation hubs mainly include steel structure for the roof and reinforced concrete structure for floors and partitions. The characteristics of large space construction include large span and great length. In order to ensure the indoor transparency, the roof system normally adopts steel frame, steel truss, and steel suspension structure. In terms of fire safety, steel structure cannot withstand high temperature; once the temperature exceeds 500 °C, the yield strength will drop drastically; consequently, the roof will collapse due to the loss of support. Therefore, it is very important to include fire safety design in the roof of steel structure.

The large space design adopted by large transportation hub buildings will use a lot of steel structure, and the main purpose of the performance-based fire safety analysis for steel structure is to determine the fire safety performance of steel structure and fire safety measures. For steel structure buildings with multiple functions, large span and large space, performance-based design methods with a scientific ground can be adopted, identifying dangerous design fire scenarios, simulating actual temperature rise in the event of fire, verifying the fire resistance performance of steel structure, and making reasonable and effective fire safety proposals. The building's waiting area has large space and relatively small fire load density, and its temperature rising in the event of fire is far different from that of small space; therefore, structural fire resistance design is done based on the temperature rise curve obtained from analysis of fire fluid dynamics.

The check-in hall on the +12.15 m level of Shanghai Hongqiao Western Terminal adopts 36 m large-span steel structure as its roof. A solid web steel beam of variable sections (small ends and large middle section) are used, with one end hinged with the main frame and the other end hinged with the steel column (Fig. 4.9).

Considering that in extreme cases there is a possibility that some luggage in the check-in hall may catch fire, the 1.5-h 8 MW steady-state fire is used as the fire scenario for design. In the designed fire scenario, the designed temperature of the steel components of the +12.15 m check-in hall roof of the Western Terminal is shown in the figure below.

The designed temperature of the roof's steel components is about 317 °C~342 °C; the steel columns and the lower parts of the bracing between the columns may be directly surrounded by the flame, and their temperature is close to that of the flame; the designed temperature of the steel columns and the parts of the bracing higher than 7 m is lower than 473 °C (Fig. 4.10).

The calculation of the steel components of the +12.15 m check-in hall roof of the Western Terminal is checked with the fire resistance bearing capacity method, and the load ratio in case of fire is obtained by calculation with SAP2000. The designed temperature and the ratio between the internal force and the bearing capacity of various components are shown in Table H8. It can be seen from the table that in the designed fire scenario, the internal force of the roof steel components is smaller than

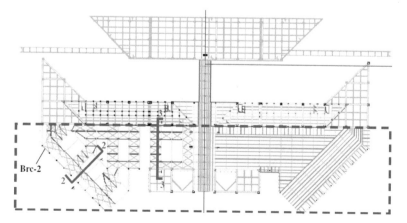

Fig. 4.9 Floor plan of the +12.15 m check-in hall roof of Shanghai Hongqiao Western Terminal

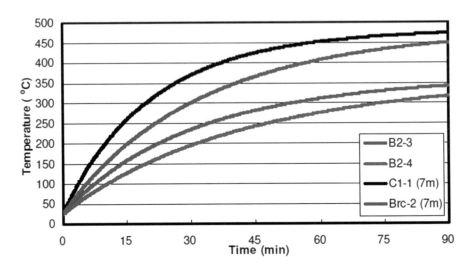

Fig. 4.10 Designed temperature of steel components of the +12.15 m check-in hall roof of the Western Terminal

Table 4.2 Internal force and bearing capacity of steel components of the +12.15 m check-in hall roof of the Western Terminal in case of fire

Name of component	Designed component temperature(°C)	Component internal force/bearing capacity in case of fire
Steel beam B2–B3	342	0.39
Steel beam B2–B4	317	0.34
Steel column C1-1	Below 7 m – Close to the temperature of flame Above 7 m – 473	–
Bracing between columns Brc-2	Below 7 m – Close to the temperature of flame Above 7 m – 450	–

their bearing capacity. Therefore, it is recommended that steel components without fire protection should be adopted for the +12.15 m check-in hall roof of the Western Terminal. The temperature of the steel columns and the parts of the bracing with a height less than 7 m is close to that of flame, so they need fire protection. Considering the 1.5 safety coefficient, the protection height for the columns and the bracing between the columns needs to reach 10.5 m. Given the entire columns' height of 12 m and based on the designed fire scenario, it is recommended that 1.5-h fire protection for full height of the columns and bracing (except for seismic-resistant components) be adopted (Table 4.2).

4.5 Safe Evacuation

4.5.1 Safe Evacuation in Transportation Hub Buildings

(1) Evacuation route

People in the middle area of the elevated waiting areas directly evacuate to the drop-off platform on the two sides or go down to the platform level via the entry staircase; people in the waiting area on the southern and northern sides go down to the platform level through the entry staircase in the middle or evacuate to the drop-off platform through the middle of the waiting area or evacuate to the first floor through the staircase of the southern station or the northern station.

The check-in gate in the elevated waiting area will automatically open in the event of fire, and people in the ancillary buildings can evacuate to the first floor through the internal staircase (Fig. 4.11).

Following is the analysis of the elevated floor using a specific case. If fire occurs on the elevated floor, people can evacuate to the base platform through the GT1–GT2 and GT13–GT14 evacuation staircases or evacuate to the middle platform through GT3–GT12. Some of the people can evacuate to the base platform through the closed staircases A1–A4 or directly evacuate to the elevated drop-off platform through GK1–GK2 (Fig. 4.12).

(2) Evacuation strategy

    1. Phased evacuation strategy

Large transportation hubs normally have large areas; there are many staircases and exits for fire evacuation, and the evacuation routes are more complicated

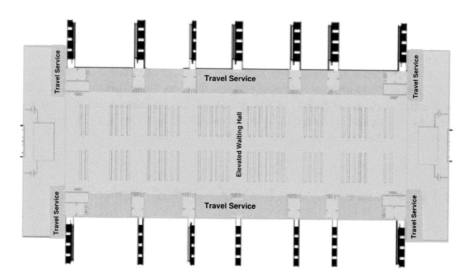

Fig. 4.11 The main functions of the elevated floor

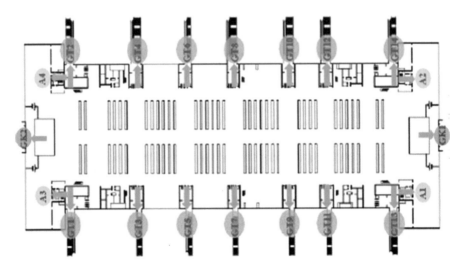

Fig. 4.12 Elevated floor evacuation

compared with other public buildings. In case of fire, a phased evacuation strategy can be adopted, evacuating people separately based on their distance to the fire area to mitigate casualties – people closer to the fire area will be urgently evacuated to place farther away from the fire or to a safe area, and people farther away from the fire will be evacuated to the final safe area.

2. Ensure the mobility of the evacuation exit and guide evacuation

Maintaining the mobility of the evacuation exit is critical for the evacuation of the people in various areas of the station; therefore, there should not be articles that obstruct the evacuation passage. No obstacles should be allowed in front of various evacuation exits, and the evacuation routes should be directly connected with the evacuation exits. In addition, it is recommended that flat-opening fare gates should be adopted, which can open together in case of fire; mobile barriers can be placed along the two sides of the check-in gate based on the actual situation, increasing the valid width of the exit in case of emergency evacuation. On the critical nodes of the evacuation route, evacuation guiding signs complying with standards should be reasonably placed, and necessary safety personnel should be added to guide the evacuation so that the evacuation time can be reduced significantly (Figs. 4.13 and 4.14).

4.5.2 Special Evacuation Design for Airport Terminals

(1) Adoption of boarding bridges as evacuation exits

In large airport terminals, the fixed connecting overpasses of the boarding bridge are part of the general strategy for fire evacuation in the terminal. On the far-end conversion platform of each fix-connected overpass away from the terminal, there is

Fig. 4.13 Flat-opening
check-in gate

Fig. 4.14 Mobile barrier

a fully open steel ladder with a width of 1.2 m–1.5 m leading to the tarmac. From the terminal's second floor departure lounge and mezzanine to the corridor, along the fix-connected overpass, these stairs can be reached separately. These fix-connected overpasses are used only for the movement of people, so the fire load is small.

The corresponding fix-connected overpass cannot be used only when passengers are boarding or disembarking or in the case of a fire in the terminal occurring near the entrance to the fix-connected overpass and the access path being obstructed. The number and distribution of boarding bridges in the corridor lounge are large, and they are evenly distributed, which ensures that people can always be evacuated to another fix-connected overpass.

Once passengers reach the fix-connected overpass, people will be in a relatively safe area. The door entering to the fix-connected overpass is a fireproof door. In the terminal of Capital Airport, the fire limit of such fireproof door is required as accurate as minute, which avoids the fire in the main building affecting boarding bridge before the completion of the evacuation (Fig. 4.15).

Fig. 4.15 Real shot of Changbei Airport boarding bridge

4.5.3 Emergency Evacuation in Transit Tunnels

(1) Evacuation route

A longitudinal evacuation route is built on one side of the rail level of the Chongming transit tunnel, which is connected with the door of the train through an opening, and the upper part is connected with the highway level through staircases for passengers to evacuate up to the road level.

The passage on the side is longitudinal, and its southern and northern ends extend to the manhole, which is directly connected with the evacuation staircase (the width of the staircase is 1.2 m) on the ground floor. The clear width of the evacuation route is not less than 1 m (in the pump room).

(2) Evacuate the train

On the side of the platform at 1:1 width, there is a doorway connected with the rail area. The size of the doorway is 1000 mm×2180 mm, and each carriage has a door corresponding with it.

In case of evacuation, it should be considered to open all the doors in the last carriage, and evacuate passengers from the doorways corresponding with the doors to the safe evacuation route.

The width of the evacuation route is the main constraint for the evacuation speed; therefore, opening the doors of the last carriage will not impact much the time of escape, but can slow down the entry of smoke into the carriage to a larger extent (Figs. 4.16 and 4.17).

(3) Ways and conditions of evacuating the accident tunnel

1. Evacuation staircases are set with an interval of 280 m along the longitudinal evacuation route, and there are 27 staircases/lines; the opening of the evacuation staircase is about 1.3 m (width) × 2 m (length), taking 0.3 m of the vehicle lane; in the landing of the staircase, there are clear signs of "No Stop."
2. When the rail level malfunctions, people need to evacuate to the first lane on the left of the upper road level or evacuate through the evacuation staircase of the manhole. According to the traffic dispersion simulation of the upper highway level, it is tentatively decided that people will evacuate to the opposite direction of the train when the accident occurs, the traffic dispersion on the upper road level will complete in 100 s, and the two evacuation cover plates closest to the tailstock of the train will be opened for vertical evacuation.

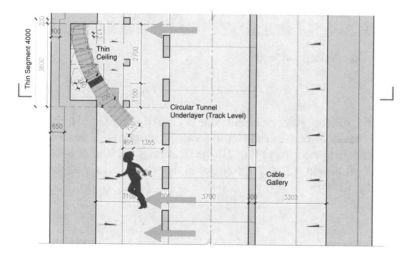

Fig. 4.16 Floor plan of the rail level

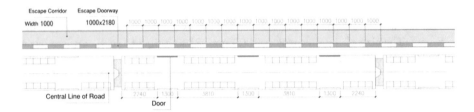

Fig. 4.17 Floor plan of the corresponding relationship among the train door, the evacuation route, and the doorway

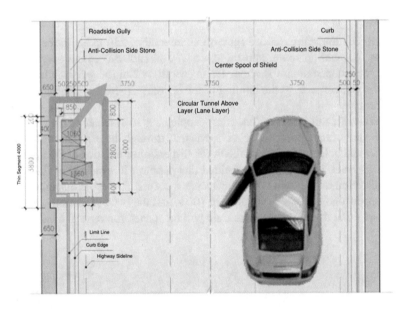

Fig. 4.18 Plane of the highway level

3. If the accident point (accident point is located at the end of train) is between the manhole and evacuation staircase 4, people will evacuate from the manhole to the ground floor; if the accident point is after evacuation staircase 4, people will wait for rescue in the waiting area after the evacuation (Fig. 4.18).

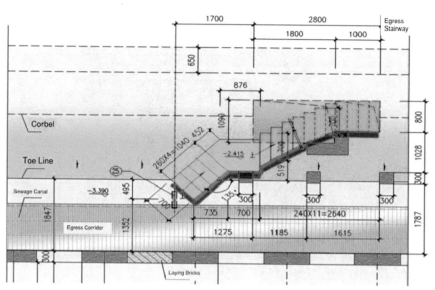

Evacuation Staircase for the Lower Rail Level

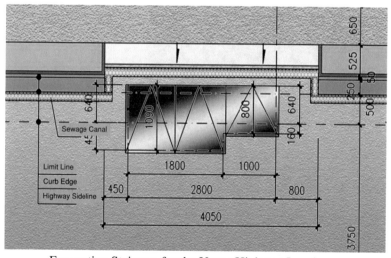

Evacuation Staircase for the Upper Highway Level

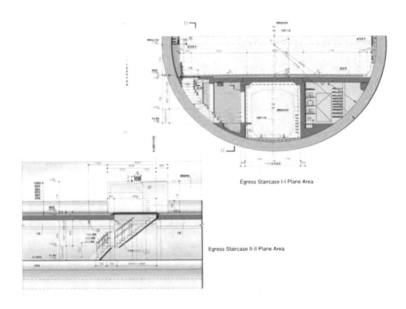

Egress Staircase I-I Plane Area

Egress Staircase II-II Plane Area

(4) Highway linkage requirement

1. Adopt traffic guidance technology: in principle, a VMS (Variable Message Sign) board is placed every 1.2 km in the tunnel. Considering the special combination of highway and light rail of the Changjiang tunnel bridge, it is recommended that a VMS is placed every 500 m.

2. Manage vehicle guidance

Arrange emergency motor vehicles to assume the responsibility of lane clearance in case of emergency. After an accident occurs, relevant vehicles will separate the first lane on the left and set up alert signs.

(5) Emergency evacuation comparison table (Table 4.3)

It can be seen from the comparison, considering the special requirements of existing tunnels, the evacuation ability of Changjiang Tunnel of Chongming line is better than general tunnels.

4.5.4   Special Evacuation Strategy for Subway Stations

In the subway space, the personnel are mainly distributed in the public area of the station hall and the platform and the train carriage. When a fire breaks out on the platform floor, passengers on the platform level, after several seconds of reaction, will reach the station hall floor via stairs and then through the station hall level to the secure exit. Since there are very few combustibles in the public area of the station hall and the closed underground stations are equipped with ventilation and smoke removal system for accident, when a fire occurs at the platform floor or in the interval tunnel, the public area of the station hall can provide a high security guarantee for the evacuation of people on the platform floor within a certain period of time. [11]

Both Fire Protection Standards for the Design of Subway (GB 51298) and Standards for Fixed Rail Transportation and Rail Passenger Transportation System (NFPA 130) of China require that guests on the platform level can evacuate the platform within 4 min to configure the number of escalator groups and total transportation

Table 4.3 Shanghai Chongming Tunnel emergency evacuation comparison table

Normal tunnel section requirements	Actual configuration of Changjiang Tunnel of Chongming line
A connection passage is set between two sections, and the smallest horizontal distance between the two neighboring connection passages is no greater than 600 m	Evacuation routes connecting with the highway level are set, and the distance between the evacuation routes is about 280 m
A longitudinal evacuation platform should be set inside the section (normally the width is 600 mm)	Longitudinal evacuation routes with an evacuation width not smaller than 1000 mm are set
A smoke-free staircase with the direct access to the ground should be built within the air shaft of the section	A smoke-free staircase with the direct access to the ground should be built within the air shaft of the section
The time it takes to evacuate to the safe area normally exceeds 1 h	Even under the working condition of serious overload (9 people/square meter), it takes about 46 min to evacuate to the safe area

capacity from the platform to the public area of the station hall or out of the ground. Besides, they also require all passengers to be evacuated to the public or other safe areas of the station hall within 6 min to match the relationship between the evacuation stairs on the platform, the setting position, the number and width of the escalators, and the height or length of the station hall or other safety zone from the platform:

$$T = \frac{Q_1 + Q_2}{0.9[A_1 (N-1) + A_2 B]} \le 4 \min$$

where:

Q_1: The number of passengers (people) on a train entering the station during the long-term or passenger flow control period for the maximum passenger flow of the super-peak hour

Q_2: The maximum number of passengers waiting on the super-peak hour platform during the long-term or passenger flow control period (persons)

A_1: Passing capacity of an escalator [person/(min·set)]

A_2: Passing capacity of evacuation stairs per unit width [person/(min·m)]

N: The number of escalators used for evacuation (set)

B: The total width of the evacuation stairs (m) (the width of each group of stairs should be calculated as an integral multiple of 0.55 m)

With urbanization process of China, rail traffic in China's major cities have been widely adopted. For example, Shanghai, Guangzhou, Beijing, Shenzhen, and other major cities have planned and constructed ten or even more rail lines, and some rail transit stations serve as a commuter station for three or even four subway lines. Part of the subway station design and construction also integrate with the humanities, attractions, geological conditions, integrated hub, and other factors with local features with a variety of architectural space forms.

In the Zhuguang Road Station of the newly built Shanghai Line 17, an atrium-style platform and station hall are adopted; in this case, whether the station hall can be regarded as equivalent safe area to adopt 6 min safe evacuation time is debatable.

The length of subway lines in Beijing ranks first globally. There are 47 commuter stations, which, according to different construction conditions, geological conditions, are constructed as point-type commuter stations with variety of spatial forms. Due to huge traffic, commuter stations often have large space scale; for example, the area of Songjiazhuang commuter station has reached nearly 20000 m², as shown in Fig. 4.19 (inner structure of Songjiazhuang Station). The Liuliqiao commuter station adopts a special circular commuter space to reduce the possibility of danger during commuting with large traffic, as shown in Fig. 4.20 (inner structure of Liuliqiao Station). Due to the limitation of construction conditions, the ground platform of the Hujialou commuter station was constructed into a ring structure, as shown in Fig. 4.21 (inner structure of Hujialou station).

For the abovementioned stations with special architectural space form, compared with single island-type station, its architectural structure, function, and characteristics

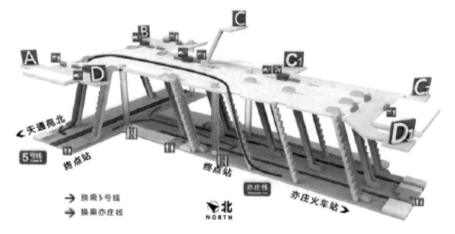

Fig. 4.19 Inner structure of Songjiazhuang Station

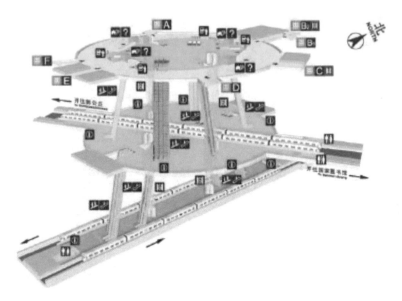

Fig. 4.20 Inner structure of the Liuliqiao Station

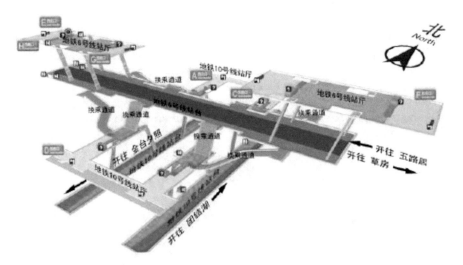

Fig. 4.21 Inner structure of the Hujialou Station

of the flow of people are particular, fire risk is greater, the spread of fire smoke is also affected by a variety of factors, and the applicability of the normative provisions of the way is still questioned. Under such circumstances, it is possible to carry out performance-based fire prevention design according to the characteristics of specific sites and fire safety objectives, and reconsider the design parameters of fire protection based on the law of smoke diffusion and the pattern of evacuation behavior, and to propose targeted requirements for the fire-fighting facilities and operating time.

egress passageway

Fig. 4.22 Layout of the first floor evacuation aisle of the GTC of an airport

4.5.5   Application of Egress Passageway in Traffic Hubs

Due to the large depth of the traffic center, part of the security exit or evacuation stairwell of the fireproof zone in the first floor cannot be straightly connected to outside; it is designed to perform evacuation through the egress passageway to the outdoor (the location of the stairwell location is shown below). If the evacuation design indeed cannot meet the requirements of Article 5.5.17 of GB50016-2014 *Fire Code for Building Design*, "Stairwells shall be directed outdoors on the first floor," the expansion of the closed stairwell or smoke-proof stairwell in the front room of the first floor can be used (Fig. 4.22).

4.6   Design of Water Fire-Fighting System for Large Transportation Hubs

4.6.1   Water Fire-Fighting System for Buildings of Transportation Hub Stations

1. Suitable automatic fire suppression systems for large spaces

   According to the analysis and comparison of the above automatic fire extinguishing methods, the automatic fixed fire water monitor system, the high water cannon device with automatic scanning and water jetting functions, and the sprinkler sys-

Fig. 4.23 Large space fire water cannon map

tem with automatic scanning and positioning functions in large spaces belong to automatic fire extinguishing systems which are suitable for the large spaces of the waiting halls in rail transit hubs (Fig. 4.23).

On this basis, considering the spatial characteristics, design aesthetics, and overall transparency effects of some elevated waiting areas in some stations, architectural structures, such as Tianjinxi Railway Station, Wuhan Railway Station, and New Guangzhou Station with arched and large curved roofs are suitable for automatic fixed fire water monitor system:

(1) An automatic fixed fire water monitor system shall be installed in the waiting hall, and the water jetting of at least one water monitor shall reach any part of the protected area.
(2) The fuel gas set up in the kitchen of the auxiliary building shall be set up against the external wall, with sufficient explosion venting surface, and automatic fire extinguishing facilities shall be set up in the kitchen.
  2. Hydrant system

The waiting hall is a large-span space. Indoor hydrants are generally arranged along the wall surface of the hall. In the middle of the hall, refrigerator hydrants or underground hydrants are arranged on the ground. Points cannot be blocked by fixed furniture and light warning signs are provided for easy access in case of fire. Fire hydrants shall also be provided at the drop-off platform outside the waiting hall.

4.6.2   Traffic Tunnel Water Fire-Fighting System

In the design of tunnel engineering, especially for super-long tunnels or tunnel groups longer than 20 km, emergency rescue stations are required to be set up in the code, and fire hydrants and high-pressure water mist extinguishing systems are required to be set up in the emergency rescue stations.

(1) Hydrant system

Taking Chongming Tunnel in Shanghai as an example, the upper and lower tunnels, the lower and upper cable tunnels, and the rail transit tunnels are all independent of each other in terms of civil structure and have strict fireproofing separation. In the water fire control design, the upper highway tunnel, the lower rail transit tunnel, the evacuation platform, and the lower cable tunnel are considered as a whole, sharing the fire water source, the fire pressurization equipment, and the hydrant system equipment. The hydrant system, the high-pressure water mist system in the cable tunnels, the fire pump room, and the fire extinguisher facilities have been built and put into use with the upper highway tunnel. At present, the hydrant in the lower evacuation platform is only equipped with pipelines and stoppers, and the corresponding supporting facilities such as pump start button and alarm button will be improved when the rail transit is constructed.

In this project, fire pump rooms are set up separately at both ends of the tunnel (Pudong end and Changxing Island end). Two municipal water supply pipes of DN250 are introduced into fire pump rooms from the municipal water supply pipe network at both ends. After the antipollution partition valve group is set up in the pump house, they are connected to each other to meet the two-way water supply demands. During fire-fighting, the booster pump directly pumps water from the connecting pipe section, without the fire pool. Each fire pump room is equipped with a hydrant pump set, a water spray pump set, and a foam pump set, and another set of high-pressure water mist pump set is arranged in the fire pump room at the Pudong end.

The water supply range of the hydrant pump set is the upper highway tunnel, the lower rail transit tunnel, and the evacuation platform; two DN150 hydrant main pipes of the fire pump rooms at both ends are connected to the lower pipe gallery of the two-hole tunnel, respectively, and penetrate through the whole line; and the two main pipes are connected with each other at the connecting passage to form a safe and reliable ring-shaped water supply main pipe. And the branch pipes of the main pipe every 50 m are connected, respectively, to the hydrant box, and hydrant stoppers are installed in the one side of the upper lanes and the lower evacuation platform.

The designed water quantity of hydrant system in rail transit section is 10 l/s, the water column filled with water gun is no less than 10 m, and the dynamic pressure of hydrant stoppers exceeds 0.5 MPa. A hydrant is arranged in the evacuation platform of each tunnel at a spacing of 50 m, corresponding to the hydrant box position of the upper lane. Because the upper lane and the lower rail transit tunnel are separated strictly from each other by fire prevention, belonging to two fire separation areas, and not considered to be used at the same time. The upper and lower hydrant boxes share the same hydrant branch pipe for water supply. One fire pump start button and one alarm button shall be arranged at each hydrant in the evacuation platform (Fig. 4.24).

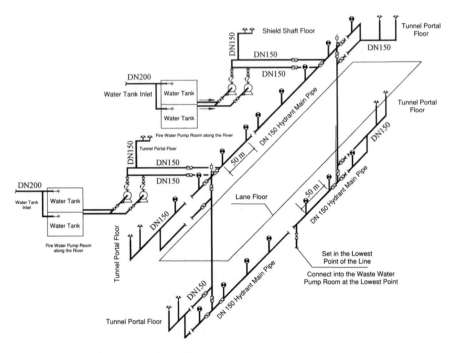

Fig. 4.24 Schematic diagram of tunnel hydrant system

(2) High-pressure water mist system

As the transformer equipment buried underground for the upper highway tunnel is locally arranged in the emergency passage of the project, the risk of electrical fire in tunnel is slightly larger than that in general subway sections. And the distance between underground sections is too long, so it is relatively difficult to find fire-fighting equipment and put fire out in time. Therefore, it is considered to add automatic high-pressure water mist local application system to the area of transformer equipment buried underground set up in groups. The specific plans are as follows: A high-pressure water mist water supply main pipe is introduced into the emergency passage, and the regional control valve is arranged beside the area of each group of transformer equipment buried underground. Water mist sprinklers are arranged on the post-valve branch around the equipment. After receiving two-way fire detection alarm signals, the water mist area control valve in the corresponding position is opened, and the local application system sprays water to extinguish or control the initial fire in the area to protect the transformer equipment buried underground and ensure the operation safety of the rail transit tunnel.

High-pressure water mist system pump set in the cable gallery can be shared by the water supply equipment of the high-pressure water mist local application system. A main pipe of high-pressure water mist local application system in emergency passage and water flow indicators, signal valves, etc. are separated from escape channel and connected separately from the pressurized pipe behind the pump set.

Fire detection and alarm facilities should be added to FAS system synchronously in the protection area, and fire alarm signals should be transmitted to the central control room of the rail transit project. Open the control valve in the corresponding area in time after confirmation, and start the high-pressure water mist pump group in conjunction with the local system spraying water to extinguish the fire.

4.7 Fire Rescue

4.7.1 Fire Rescue for Transportation Hub Buildings

(1) External rescue

External rescue means setting up fire lanes and fire-fighting sites around the building. For example, the following measures are included in designing external fire rescue for Shanghai Hongqiao Transportation Hub (Fig. 4.25):

- Fire engines have direct access to the ±0.00 m level and +12.15 m level.
- Fire-fighter can access the roof on the +24.15 m level using the fire-fighting stairway set up on the exterior wall on the +12.15 m level of the building.
- +24.15 m roof large platform with fire hydrant for fire-fighters.

(2) Internal rescue

In addition to conducting external rescue, fire-fighters also need to enter the building and carry out fire rescue in the fire scene. The stairway allows fire-fighters to enter the higher levels with fire-fighting equipment to put out the fire and rescue and evacuate the injured or elderly and weak or disabled people. It avoids the "clash" between fire-fighters and evacuated people on the evacuation staircase, which will delay fire-fighting and affect people evacuation; meanwhile, it prevents fire-fighters

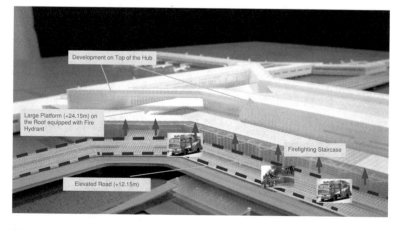

Fig. 4.25 Shanghai Hongqiao Transportation Hub external rescue illustration

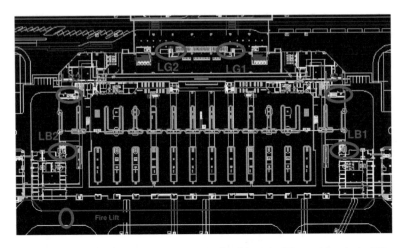

Fig. 4.26 Location of the ±0.000 m main building fire lift of the Western Terminal of Hongqiao Transportation Hub

from consuming too much of their energy by climbing the staircase and as a result cannot start fire-fighting quickly. Fire lifts of the Western Terminal of Shanghai Hongqiao Transportation Hub are shown as follows (Fig. 4.26):

4.7.2   Fire Rescue in the Traffic Tunnel

Take the fire rescue of Chongming Tunnel as an example:

(1) Integrated bridge-tunnel management
    Implement integrated management of the tunnel, bridge, and ground road, and centralize monitor, management, maintenance, and emergency response.

(2) Management center
    Changxing Island Management and Control Center – Central control room, emergency vehicle, and equipment
    Pudong Management Subcenter – Monitor duty room, emergency vehicle, and equipment

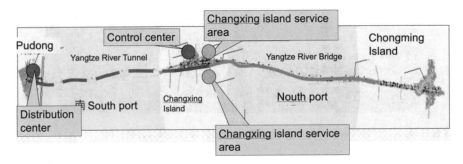

1. In case fire occurs in the rail section, the highway and rail in the tunnel will carry out fire-fighting and rescue under the central command of the central control room.
2. The tunnel section will start alarm and smoke exhaust. On the upper highway level, entry of vehicles will be controlled, and vehicles inside the tunnel will slow down and evacuate outside of the tunnel while keeping an emergency access.
3. Passengers on the train within the section will leave from the side door and proceed to the evacuation passage, where they go to the staircase landing along the longitudinal direction, and arrive at the upper highway level.
4. Rescue personnel enter the tunnel through the highway level, and arrive at the rail session through the staircase to implement rescue.

4.8   Fire Safety Management

The unique scale and nature of large transportation hub projects bring many challenges. In particular, a large people flow inside the hub need to be controlled through complicated safety measures. Considering these factors and the emergency evacuation that may be triggered by possible accidents, the entire hub required high-level collaboration and control capabilities so as to safely evacuate all the people.

The management measures of large transportation hubs should include a complete set of effective fire safety management system, which plays a critical role in maintaining the existing fire safety standards. The management tasks include:

- Establish a complete set of hub security and evacuation strategies.
- Educate and train staff and tenants so that they master the necessary skills of handling fire emergencies. The training should include evacuation steps and the use of portable fire-fighting devices.
- Monitor and check areas where fires are easy to occur.
- Develop a maintenance plan to ensure the liability of fire-fighting devices.
- Examine changes in the building and/or fire-fighting safety equipment, and make sure that they conform to fire-fighting safety strategies and other related standards.
- Examine passenger forecast and data to estimate possible impact on evacuation facilities.

4.8.1   Routine Maintenance and Care

Large transportation hubs are buildings with high people density, and it is very important to ensure that the risk of fire is reasonably minimized. More importantly, we must ensure that all the escape routes are usable in case of fire. Therefore, stringent routine maintenance and care is critical. For this purpose, the following considerations should be included in the management strategies:

- Implement strict management of the combustible substances in various areas inside the composite hub, and timely eliminate such substances to reduce the first risk.
- Piling of rubbish or combustible substances in public areas or evacuation routes without permission is not allowed.
- A large amount of waste will be generated inside the hub. Measures should be taken to remove such waste from the building to prevent piling. It can be placed in a safe storage area first before it is eventually shipped out.
- Staff and tenants should enhance regular housekeeping and equipment maintenance to maintain sound fire safety standards.
- Tenants should ensure that all their commodities are put inside the boundary lines of their shops to prevent fire and smoke from spreading to areas outside of the retail shop. If the retail store is equipped with fire-resistant shutter, it needs to be ensured that the opening and closing of the shutter are not obstructed by commodities or other stuff.

4.8.2 Staff Training

Staff, security guards, and policemen will conduct nonstop patrol inside the hub so that any accident can be spotted at the early stage. In case of fire, staff should immediately adopt relevant response measures. For this reason, relevant training should be provided for staff, teaching them how to take emergency measures, including raising the alarm, warning the public, and using portable fire extinguishers. In addition, special training should be provided for staff responsible for guiding and assisting the evacuation of people.

4.8.3 Fire-Fighting Equipment

For small fires, it is likely that people in the fire scene can put out the fire successfully, so there is no need to evacuate most of the areas inside the hub. Various areas in the hub should have portable fire extinguishers and fire hose reels for fast fire extinguishing.

4.8.4 Control and Communication

Large transportation hubs should have their own fire control rooms, in which all the fire safety and security systems of all areas can be operated.

Systems within the fire control room that are related to fire safety should include the communication system, alert display board, CCTV monitor, fire alarm voice broadcasting control, ventilation and smoke extraction system, and other detection, alarm, and fire extinguishing devices, such as automatic sprinkler systems.

The fire control room should have trained staff on duty uninterruptedly.

An effective communication system should be provided for use between the staff responsible for evacuation and the control room.

The CCTV camera system should cover the entire hub. As an auxiliary fire detection system, CCTV is also one of the important tools to monitor the evacuation process. CCTV monitors should be installed inside the control room so as to closely monitor the fire and the evacuation process.

4.8.5 Assistance in Evacuation

In case of emergency evacuation, mobility-impaired people inside the transportation hub need special help. This introduces some safety characteristics, and staff responsible for evacuation should understand these safety characteristics.

- The evacuation strategy allows people to evacuate horizontally to the neighboring areas inside the building.
- Trained staff must be in place to spot and help mobility-impaired people to evacuate. Mobility-impaired people include wheelchair users, crutch users, and elderly people.
- Staff also need to be trained to help vision-impaired and hearing-impaired people to evacuate.

4.9 Summary of the Chapter

The first part of this chapter presents six typical topics in fire safety design of rail transportation hubs according to their construction characteristics, fire compartmentation, smoke compartmentation, evacuation distance, safety evacuation, large space steel structure fire safety, and commercial facilities, and then introduces some important concepts of performance-based design, together with some important performance-based evacuation assessment parameters.

The second part analyzes fire safety difficulties in the four typical spaces of railway integrated transportation hubs, namely, the tall and large composite space, vertical through space of the atrium, underground space, and dense space, and proposes fire safety strategies in the three areas of building fire safety, safe evacuation, and fire rescue.

Among them active fire safety strategies include four aspects, fire resistance rating, fire compartmentation, fire separation, and rescue conditions, and passive fire safety strategies include five aspects covering fire water supply, smoke control system, automatic fire alarm system, fire safety equipment preparation, and evacuation guidance system.

Chapter 5
Case Analysis of Performance-Based Fire Safety Design for Large Transportation Hub: Shanghai Hongqiao Integrated Transportation Hub

5.1 Project Overview

5.1.1 Overview

As the largest modern transportation hub project in the Yangtze River Delta constructed in Shanghai, Hongqiao Integrated Transportation Hub integrates various transportation modes including air, railway, highway passenger transportation, and public traffic transfer. After completion, the hub will connect the three metropolitan circles of Shanghai, Nanjing, and Hangzhou as well as economically developed areas like Suzhou, Wuxi, and Changzhou, covering a development network with Shanghai as the central economic city.

The project is located to the west of Shanghai's central urban area, covering the Outer Ring to the east, the Outer Ring Line of the existing railway to the west, Beicui Road and Beiqing Highway to the north, and Huqingping Expressway to the south. The total project area is about 26.26 square kilometers, with a planned core area of about 625,511 m². From the east to the west are the Western Terminal, the Eastern Transportation Square, the Maglev Station, the Beijing-Shanghai High-Speed Station, and the Western Transportation Square (see Fig. 5.1). The performance-based fire safety design contained in this case only involves the Western Terminal [13].

In order to embody the design concepts of people orientation and smooth flow, sound horizontal and vertical interconnectivity and tall and large space are the main construction characteristics of the hub. Three major traffic transfer levels of −7.95 m/−9.35 m, +4.20 m/+6.60 m/+7.65 m, and +12.15 m are built for the hub (see Fig. 5.2).

© Springer Nature Switzerland AG 2021
F. Li, H. Li, *Fire Protection Engineering Applications for Large Transportation Systems in China*, https://doi.org/10.1007/978-3-030-58369-9_5

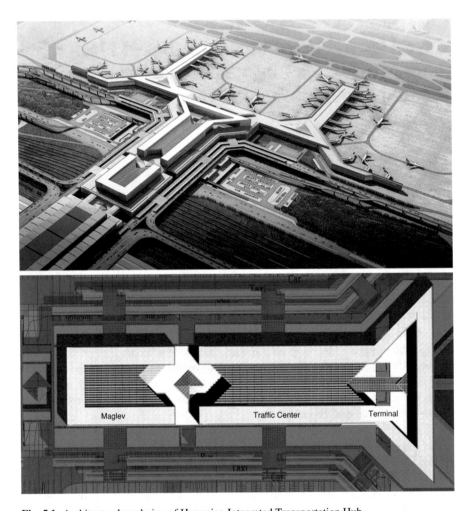

Fig. 5.1 Architectural rendering of Hongqiao Integrated Transportation Hub

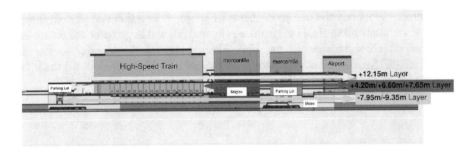

Fig. 5.2 Three major traffic transfer levels of Hongqiao Integrated Transportation Hub

5.1.2   Floor Planning

The Western Terminal, Eastern Transportation Square, and the Maglev Station are interconnected on the −7.95 m/−9.35 m level and the +12.15 m level and are connected with the +4.20 m/+6.60 m/+7.65 m level through the outdoor overpass; on the ±0.00 m level, the three sections are separated by outdoor lanes (see following figures). The Western Terminal consists of the main building and airside concourses, including the check-in hall, security inspection hall, luggage claim and processing hall, and departure and arrival concourses. Office buildings are constructed on the top of the entire Western Terminal. The Eastern Transportation Square includes five floors of garages, metro stations, and the passage levels, and ancillary commercial facilities are built on the top of the entire Eastern Transportation Square. The Maglev Station includes the underground hall, platform, exit mezzanine, and elated entry hall, and an office building is constructed on the top of the entire Maglev Station (Fig. 5.3).

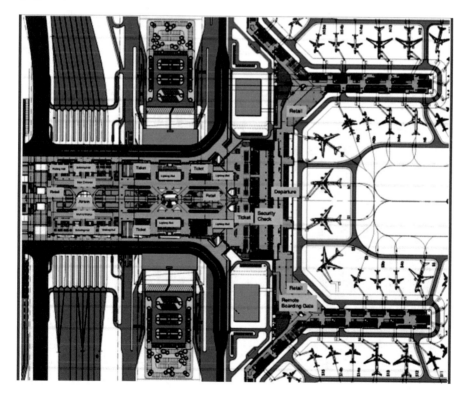

Western Terminal, Eastern Transportation Square, and Maglev Station on the **+12.15 m** level

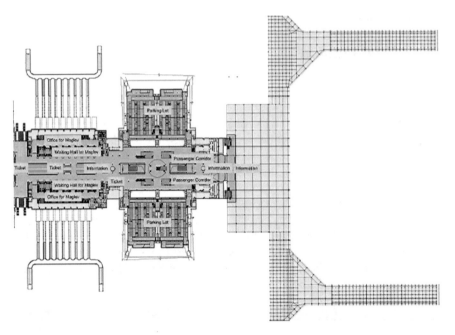

Fig. 5.3 Western Terminal, Eastern Transportation Square, and Maglev Station on the −7.95 m/−9.35 m level

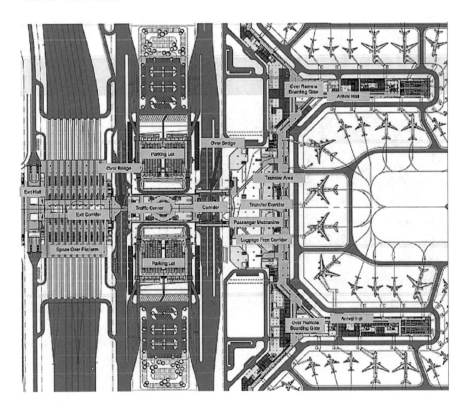

Western Terminal, Eastern Transportation Square, and Maglev Station on the **+4.20 m/+6.60 m/+7.65 m** level

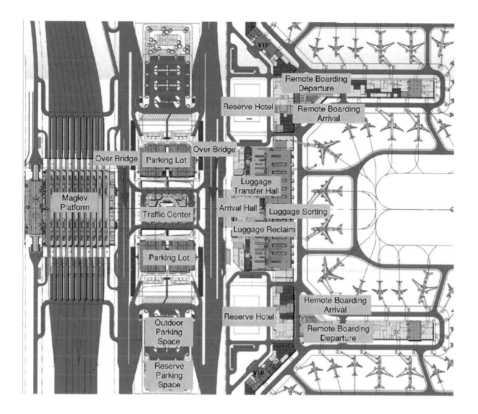

Western Terminal, Eastern Transportation Square, and Maglev Station on the **±0.00 m** level

5.1.3   Major Fire Safety Issues

Hongqiao Integrated Transportation Hub features smooth interconnectivity and tall and large spaces, making it difficult to satisfy and completely implement the existing national and local fire safety rules in areas like the division of fire safety areas, control of fire spreading, control of smoke, people evacuation strategies, and the setup of passive fire-fighting facilities, and there are also issues that need to be addressed in the absence of standards to follow, such as the division of fire safety areas in the Western Terminal.

5.2 Fire Safety Strategies

5.2.1 Fire Compartmentation Strategy

According to the use requirements and construction characteristics of the airport's check-in hall and the metro station hall, the passage between the −7.95 m self-service check-in hall and the −9.35 m metro station hall is protected by adopting the concept of fire barrier belt. According to the calculation, the width of the fire barrier belt needs to be 8 m, and the actual width is 18 m, which is used together with three light wells with a total area of 250 m². On both sides of the fire barrier belt, there are smoke-proof vertical walls that will drop together in case of fire. In addition, independent automatic sprinkler systems and mechanical smoke extraction systems are installed within the barrier belt.

The entire self-service check-in hall is a fire protection compartment, and the equipment rooms within the area are set up as independent fire separation units according to standards (Fig. 5.4).

There are many passengers in luggage claim hall, so the primary characteristic of its function is maintaining smooth flow of people, which does not allow the installation of physical division of fire compartments. Also, as there are many luggage

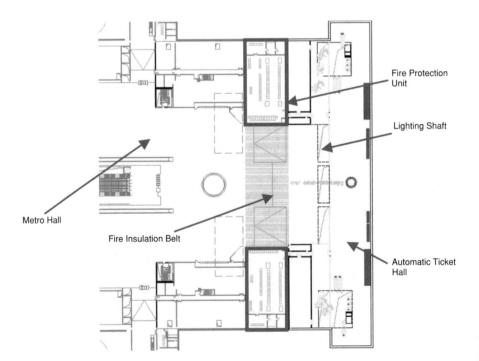

Fig. 5.4 Fire barrier belt on the −7.95 m level at the Western Terminal

transmission systems in the luggage processing hall, it is impossible to implement physical fire separation measures.

According to the characteristics of the luggage claim hall and the luggage processing hall, the following fire compartmentation measures will be adopted for fire safety.

Luggage carousels in the luggage claim hall will scatter, and there is a certain distance between carousels (about 14 m), which can effectively stop the fire from spreading. Even if there is a fire, it is difficult for it to spread within the luggage claim hall and even more difficult for it to spread to other areas with low fire load that are used as public traffic passages and have automatic sprinkler systems.

In some areas like the coffee shop within the luggage claim hall, the load of combustible substances like chairs need to be strictly controlled and should adopt Class A (non-combustible) or Class B1 (hard combustible) materials (materials should conform to relevant national inspection standards).

A fire wall of 3-h fireproof limit is used to separate the luggage processing hall and the luggage claim hall/airside concourse area. The luggage transmission system that runs across the ceiling is recommended to be wrapped with fireproof sheet to prevent fire from spreading through luggage carousels (see Fig. 5.5). It is also recommended that the luggage transmission shaft should be fireproof for at least 2 h.

In the airside concourse area, according to standards, the area of each fire compartmentation should not exceed 5000 m^2.

The arrival mezzanine, the flight connection area, and the arrival concourse are all areas clustered with people. The mezzanine is located right on top of the luggage claim hall and is connected with the airside concourse, and it is recommended to separate the two functional areas with fire resistance shutters.

Equipment rooms and ancillary rooms within the flight connection area and the airside concourse are designed according to standards. There are shops in the flight connection area, which have relatively high fire load, and the concept of "fire bulkhead" is recommended for their protection. No more physical fire separation measures will be installed in the entire circulation area (Fig. 5.6).

Similar to the departure concourse of many airports, due to the function and the large space, it is not allowed and not possible to set up extra physical fire separation measures to have horizontal division of fire safety areas within the departure concourse. In order to adapt to the architectural design concepts and functional requirements, fire compartmentation will be achieved through the following measures:

- The concept of "fire bulkhead" will be adopted for the fire safety system design for commercial areas with roof (mainly located in the corner areas of the concourse) inside the concourse, protecting and limiting these areas with high fire load and preventing fire and smoke from spreading to the large space. The size of the "fire bulkhead" should not exceed 300 m^2.
- For multiple open commercial areas within the concourse and in the +13.15 m mezzanine at the end of the concourse, the concept of "fuel island" will be adopted, strictly limiting the area, quantity, and spacing of the commercial areas directly exposed in the large space. The size of the "fuel island" is controlled at 9 m^2 and the size of the fire at 4.5 MW.

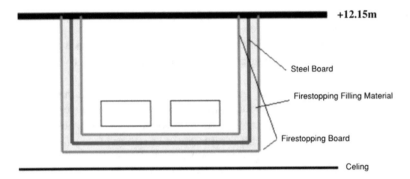

Fig. 5.5 Fire safety strategies of luggage transmission systems inside the Western Terminal

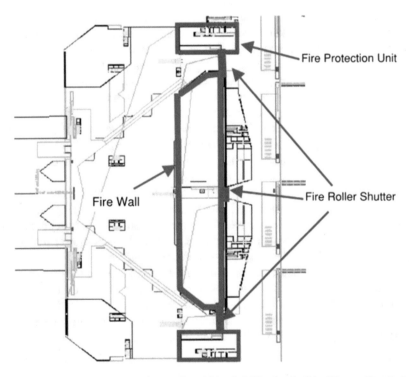

Fig. 5.6 Fire safety compartmentation on the +4.20 m/+5.25 m level of the Western Terminal

- Facilities like chairs within the concourse and coffee shops on the +13.15 m level should adopt Class A or Class B1 materials (materials should conform to relevant national inspection standards).

Due to the functional and architectural characteristics of the check-in hall and the security inspection hall, such as horizontal mobility and comfortable large space, it is not allowed and not possible to set up extra physical fire separation measures to have horizontal division of fire safety areas within the halls. According to the architectural form and functional use of the check-in and security inspection halls, in order to prevent fires within the halls from spreading to other areas without installing extra physical fire compartmentation, the following measures need to be taken:

- Install a 10 m wide fire barrier belt between the Western Terminal and the Eastern Transportation Square, and strictly forbid any combustible substances within the scope of the fire barrier belt. There are four ventilation and light wells in the area, and it is recommended that natural smoke extraction mode be adopted in combination with the light wells.
- Use fire resistance shutter to separate the passages between the check-in hall and the security inspection hall as well as the VIP lounges, and the office area is a separate fire compartment.
- The fire safety system for the VIP lounge within the check-in hall is designed with the concept of the "fire bulkhead," protecting and limiting these areas with high fire load and preventing fire and smoke from spreading to the large space. The size of the "fire bulkhead" should not exceed 300 m².
- The load of combustible substances for facilities like coffee shops within the check-in hall and chairs of the public traffic flow area should be strictly controlled, and Class A or Class B1 materials (materials should conform to relevant national inspection standards) should be adopted.
- A fire safety system should be installed where the luggage transmission system crosses the floor of the check-in hall, and the hollow partition wall above the floor should have fire-resistant rating of not less than 2.0 h. A Class A fire door that can automatically close in case of fire should be installed at the opening on the wall, and the in the upper part within the hollow wall an automatic fire extinguishing system and smoke detection system should be installed.
- The CIP/VIP leisure platform on the +13.55 m level after security inspection is protected by an automatic sprinkler system and an automatic fire detection system. Fixed furniture and equipment within the area should adopt Class A or Class B1 materials to limit the size of the fire. For large sofas that cannot adopt Class A or Class B1 materials, the concept of "fuel island" should be adopted for protection (Figs. 5.7 and 5.8).

Each floor of the office mezzanine on the +17.15 and +20.65 m level within the check-in hall is divided into two separate fire compartments. The roof restaurant is a fire compartment, and the kitchen is set up as a separate fire separation unit.

5.2.2 Smoke Control Strategy

A mechanical smoke extraction system is installed in the entire self-service check-in hall. Inside the hall, vertical smoke-resistant walls of not less than 0.5 m are used to form four smoke-proof compartments, the size of which does not exceed

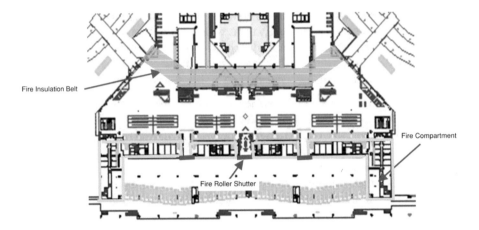

Fig. 5.7 Fire barrier belt on the +12.15 level at Western Terminal

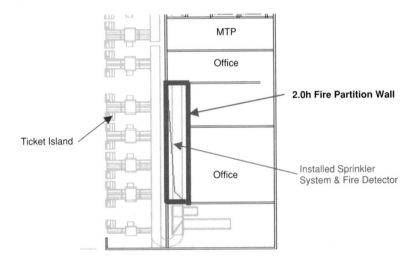

Fig. 5.8 The vertical fire compartmentation at where the luggage transmission system crosses the floor on the +12.15 m level at Western Terminal

2000 m²each, and the length of which does not exceed 60 m each. The four smoke-proof compartments share a set of exhaust fans, and the air volume of the exhaust fan is not less than 32,040 m³/h (Fig. 5.9).

A mechanical smoke exhaust system is installed in the luggage claim hall on the ±0.00 m level at the Western Terminal. Inside the hall, vertical smoke-resistant walls of no less than 0.5 m are used to form eight smoke-proof compartments, the size of which does not exceed 2000 m². It is recommended that four smoke-proof compartments share a set of exhaust fans, and the air volume of the set is not less than 103,680 m³/h, and there should not be less than nine exhaust ports (Fig. 5.10).

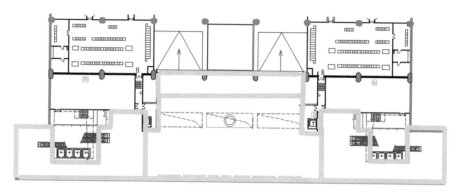

Fig. 5.9 Smoke-proof compartments on the −7.95 m level at Western Terminal

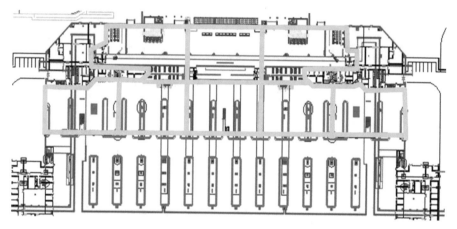

Fig. 5.10 Smoke-proof compartments in the luggage claim hall on the ±0.00 m level at Western Terminal

A mechanical smoke exhaust system is installed in the flight connection area on the +4.20 m level. Inside the hall, vertical smoke-resistant walls of no less than 0.5 m are used to form five smoke-proof compartments, the size of which does not exceed 2000 m². It is recommended that two sets of exhaust fans shall handle smoke extraction for five smoke-proof compartments, and the air volume of each set is no less than 17,640 m³/h, and there should be no less than two exhaust ports (Fig. 5.11).

A mechanical smoke exhaust system is installed in the departure concourse on the +8.55 m level. Inside the hall, vertical smoke-resistant walls of not less than 0.5 m are used to form several smoke-proof compartments, the area of which does not exceed 2000 m². It is recommended that the quantity of smoke-proof compartments that share one set of exhaust fans shall not exceed three, and the air volume of each set is no less than 90,360 m³/h (Fig. 5.12).

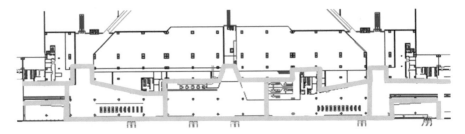

Fig. 5.11 Smoke-proof compartments in the flight connection area on the +4.20 m level at Western Terminal

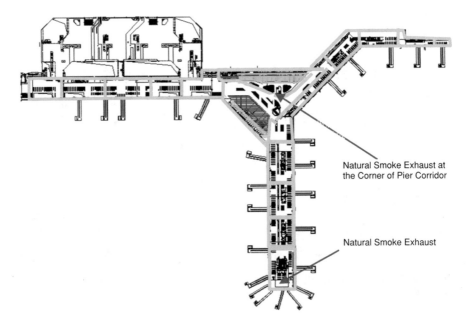

Fig. 5.12 Smoke-proof compartments in the departure concourse on the +8.55 m level at Western Terminal

Natural smoke exhaust is adopted in the check-in hall. The size of the natural smoke exhaust port is 315 m². The natural smoke exhaust effect is proven using the FDS simulation. Luggage fire is considered in the check-in hall, and the scale is 4.4 MW fast time square fire, axisymmetric plume. The purpose of the simulation is to predict the fire scene environment for the credible, dangerous fire that occurs in the section under the action of the natural smoke exhaust system, in particular the spread of heat and smoke, so as to prove the effectiveness of the natural smoke exhaust strategy (Fig. 5.13).

Fig. 5.13 Smoke-proof compartments in the security inspection hall on the +12.25 m level at Western Terminal

5.2.3 Safe Evacuation Strategy

The evacuation routes and evacuation/safety exits of the entire self-service check-in hall on the −7.95 m level are quite clear as follows:

- Two 4.8 m wide doors are for evacuation to the metro station hall to the west.
- Two 1.2 m wide evacuation staircases at the two ends on the south and the north are for evacuation to the ±0.00 m ground floor.
- The evacuation distance of the entire area is controlled within 60 m (Fig. 5.14).

The east-west span of the ±0.00 m level is about 500 m, and the south-north span is as wide as 1500 m. Evacuation together will cause operational chaos and is not necessary. Therefore, on this level, a phased evacuation strategy is adopted. The specific zoning strategy is shown in the figure below (Fig. 5.15).

Area C1 is mainly the luggage claim hall. As the +5.25 m mezzanine above it is all within the large space, once fire occurs in the area or on the +5.25 m mezzanine, the broadcasting system should notify people in this area and on the +5.25 m mezzanine, and the two areas will evacuate at the same time. People in this area can evacuate by the following means:

- Directly evacuate to outdoor through the two 4 m wide and two 2 m wide entry/exit doors on the west.
- Directly evacuate to the outdoor through the two 3.6 m wide emergency evacuation doors on the southern part and the northern part of the hall.
- Evacuate to the check-in hall on the +12.15 m level through the two 1.3 m wide staircases in the middle, and then evacuate to the outdoor. The +12.15 m level can be considered as a quasi-safe area.
- The largest evacuation distance is controlled within 60 m.

Area C2 is the luggage processing hall, where people are mainly airport staff who are familiar with the environment and can directly evacuate to the outdoor through the exit of the luggage processing hall.

Area C3 and Area C4 are mainly offices and various professional machine rooms, where people are mainly airport staff who are familiar with the environment and can

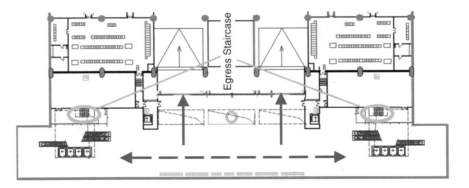

Fig. 5.14 Evacuation on the −7.95 m level at Western Terminal

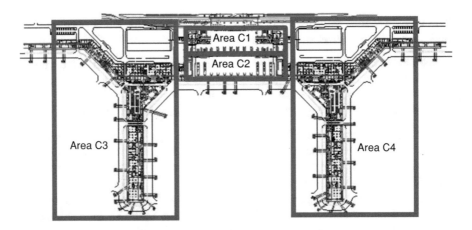

Fig. 5.15 Evacuation zoning on the ±0.00 m level at Western Terminal

directly evacuate to the outdoor through the exit of their respective fire compart-ments (Fig. 5.16).

The east-west span of the +4.20 m level is about 500 m, and the south-north span is as wide as 1500 m. Evacuation together will cause operational chaos and is not necessary. Therefore, on this level, a phased evacuation strategy is adopted. The evacuation zoning strategy is shown in the figure below.

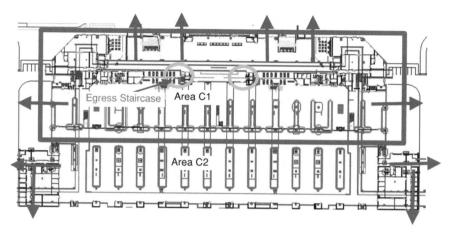

Fig. 5.16 Evacuation in Areas C1 and C2 on the ±0.00 m level at Western Terminal

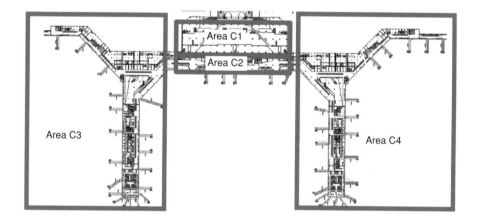

As the +5.25 m level is the large space mezzanine of Area C1 on the ±0.00 m level, once fire occurs in the area or in Area C1 on the±0.00 m level, the broadcasting system should notify people in this area and in Area C1 on the ±0.00 m mezzanine, and the two areas will evacuate at the same time. People in this area can evacuate by the following means:

- Evacuate to the transfer passage on the +6.60 m level in the Eastern Transportation Square through two 4.0 m wide access doors, and then evacuate to the outdoor.
- The maximum evacuation distance is controlled within 60 m.

If fire occurs in Area C2, the broadcasting system only notifies people in the area, who can evacuate by the following means:

- Evacuate to the outdoor apron through five 1.4 m wide boarding bridges.
- Evacuate to the ±0.00 m level through two 1.3 m wide staircases, and then evacuate to the outdoor.

- The maximum evacuation distance is controlled within 60 m.

If fire occurs in Area C3 or C4, people in Area C3 or C4 will evacuate together, and the broadcasting system only notifies people in the area where there is a fire, who can evacuate by the following means:

- Evacuate to the outdoor apron on the ±0.00 m level at the Western Terminal through 20 1.4 m wide boarding bridges.
- Evacuate to the ±0.00 m level at the Western Terminal through 11 1.3 m wide staircases, and then evacuate to the outdoor.
- The maximum evacuation distance is controlled within 60 m. Boarding bridges are used as the main evacuation routes in the entire airside concourses (Figs. 5.17 and 5.18).

The +8.55 m level at the Western Terminal is similar to the +4.20 level, with an east-west span of about 500 m and a south-north span of 1500 m. When fire occurs, it is recommended to adopt a phased evacuation strategy. The evacuation zoning strategy is shown in the figure below (Fig. 5.19).

Above Area C1 is the security inspection hall on the +12.15 m level. Once fire occurs in C1, the broadcasting system should notify people in this area as well as in the check-in hall and the security inspection hall on the +12.15 m level, and both areas will evacuate at the same time. People in Area C1 can evacuate by the following means:

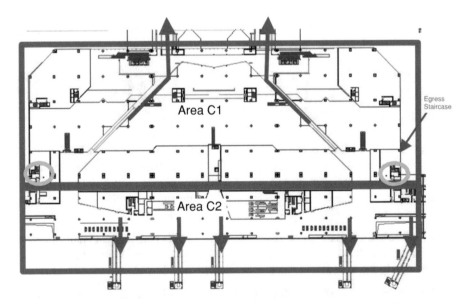

Fig. 5.17 Evacuation in Areas C1 and C2 on the +4.20 m/+5.55 m level at Western Terminal

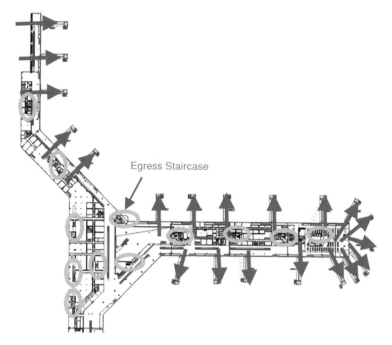

Fig. 5.18 Evacuation in Areas C3 and C4 on the +4.20 m level at Western Terminal

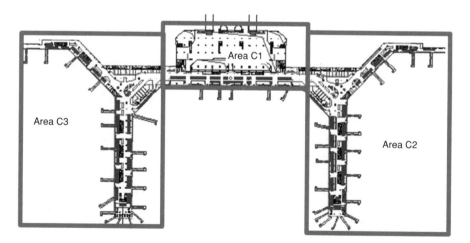

Fig. 5.19 Evacuation on the +8.55 m level at Western Terminal

- Evacuate to the boarding bridge on the +4.20 m level through five 1.4 m wide ramps, and then continue to evacuate to the outdoor apron area on the ±0.00 m level at the Western Terminal.
- Evacuate to the ±0.00 m level at the Western Terminal through two 1.3 m wide staircases, and then evacuate to the outdoor.
- The maximum evacuation distance is controlled within 60 m.

Once fire occurs in Areas C2 and C3, the broadcasting system should notify people in this area as well as in the CIP and VIP areas on the +13.55 m level, and both areas will evacuate at the same time. People in this area can evacuate by the following means:

- Evacuate to the boarding bridge on the +4.20 m level through 16 1.4 m wide ramps, and then continue to evacuate to the outdoor apron area on the ±0.00 m level at the Western Terminal.
- Evacuate to the ±0.00 m level at the Western Terminal through 11 1.3 m wide staircases, and then evacuate to the outdoor.
- The maximum evacuation distance is controlled within 60 m. Boarding bridges are used as the main evacuation routes in the entire airside concourses (Figs. 5.20 and 5.21).

The main building area on the +12.15 level at the Western Terminal is mainly the check-in hall and the security inspection hall. On the south and the north of the main building area are, respectively, CIP and VIP waiting halls. As the main building area has a completely different function from that of the CIP and VIP waiting halls and there is a certain distance between the two, evacuation together will cause operational chaos and is not necessary. Therefore, on this level, a phased evacuation strategy is adopted. The evacuation zoning strategy is shown in the figure below (Fig. 5.22).

People in Area C1 and the office area on the +17.15 m level and +20.65 m level will evacuate at the same time, and people in Area C1 can evacuate by the following means:

People in this area who have not gone through security inspection will evacuate by the following means:

- Evacuate to the outdoor elevated highway on the +12.15 m level through eight 2 m wide entry/exit gates.
- Evacuate to the transfer passage on the +12.15 level in the Eastern Transportation Square through five 2 m wide doors, and then evacuate to the outdoor elevated highway.
- Evacuate to the ±0.00 m level at the Western Terminal through four 1.3 m wide staircases in the check-in hall, and continue to evacuate to the outdoor.

The maximum evacuation distance is controlled within 60 m. People in this area who have gone through security inspection will evacuate by the following means:

- Evacuate to the +8.55 m level at the Western Terminal through two 1.3 m wide open staircases on the east, and continue to evacuate.

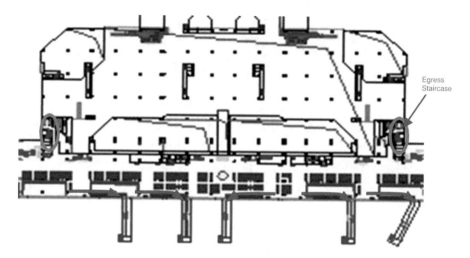

Fig. 5.20 Evacuation in Area C1 on the +8.55 m level at Western Terminal

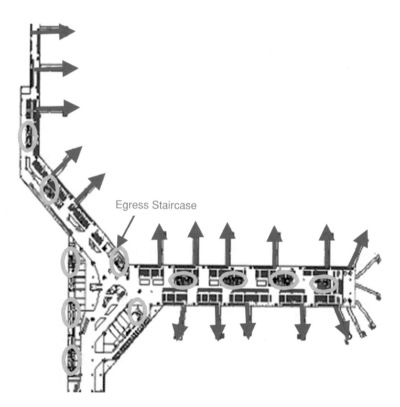

Fig. 5.21 Evacuation in Areas C2 and C3 on the +8.55 m level at Western Terminal

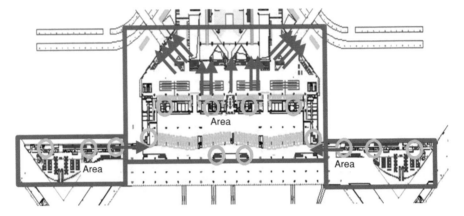

Fig. 5.22 Evacuation on the +12.15 m level at Western Terminal

- Evacuate to the affiliated office lobby of the luggage processing hall on the ±0.00 m level through two staircases in the office area on the south and the north, and continue to evacuate to the outdoor.
- The maximum evacuation distance is controlled within 60 m.

As the security inspection hall in this area is located within the large space of Area C1 on the +8.55 level, once fire occurs in Area C1 on the +8.55 level, the broadcasting system should notify people in Area C1 on the +8.55 level and in this area, and the two areas will evacuate at the same time.

People in Area C2 and Area C3 can evacuate by the following means:

- Evacuate to Area C1 on the +12.15 m level at the Western Terminal through a ramp, and continue to evacuate to the outdoor elevated highway.
- Evacuate to the ±0.00 m level at the Western Terminal through three staircases, and continue to evacuate to the outdoor.
- The maximum evacuation distance is controlled within 60 m.

Office workers evacuate to the ±0.00 m level, and then evacuate to the outdoor. As the office level is the mezzanine of the check-in hall on the +12.15 m level, once fire occurs in Area A on the +12.15 m level, the office mezzanine will evacuate at the same time. The maximum evacuation distance is controlled within 40 m.

People in the roof restaurant can evacuate to the office area on the +20.65 m level through the 1.8 m wide corridor connected with the office building, and continue to evacuate depending on the actual situation; or directly evacuate to the roof on the +20.65 m level through four 1.8 m wide gates on the south and north of the restaurant or three 1.8 m wide exit on the east. The evacuation distance is controlled within 30 m.

5.2.4   Steel Structure Fire Safety Strategy

According to analysis, the corridor and suspender steel structure on the +4.20 ~ +5.25 m level at the Western Terminal is recommended to have 1.5 h fire protection.

According to analysis, steel components without fire protection can be used for the roof the check-in hall on the +12.15 m level at the Western Terminal, while the supporting steel columns and diagonal brace for the roof are recommended to have 1.5 h fire protection.

According to analysis, the corridor steel structure on the +20.65 m level is recommended to have 1.5 h fire protection, and the steel components without fire protection are recommended for the corridor roof on the +20.65 m level.

According to analysis, the steel components for the roof of the CIP/VIP area on the +13.55 m level are recommended to have 1.5 h fire protection.

According to initial analysis, steel components without fire protection are recommended to be used for the corridor steel structure on the +36.65 m level (part of the coffee shop area at the two ends of the corridor need to adopt Class A non-combustible materials).

5.2.5   Fire Safety System

The fire detection alarm system, automatic sprinkler system, indoor fire hydrant and hose reel system, mobile fire extinguisher, evacuation guidance system, and emergency lighting system in the self-service check-in hall on the −7.95 m level should be installed according to relevant standards.

The fire detection alarm system, automatic sprinkler system, indoor fire hydrant and hose reel system, mobile fire extinguisher, evacuation guidance system, and emergency lighting system on the ±0.00 m level should be installed according to relevant standards.

It is recommended that automatic sprinkler system and automatic smoke detection system be install for relatively important luggage transmission shaft and smoke exhaust fan be installed on the top of the luggage shaft.

The fire detection alarm system, automatic sprinkler system, indoor fire hydrant and hose reel system, mobile fire extinguisher, evacuation guidance system, and emergency lighting system on the +4.20 m/+5.25 m level should be installed according to relevant standards.

It should be noted that some boarding ramps are connected with several through holes within the concourses on the +4.20 m level and the +8.55 m level, and above the ramps on the +8.55 level is a beam of exposed concrete, so adding a sprinkler system will affect the shape of the architecture. Considering that normally there are no people or fixed combustibles on the ramp, which is similar to a boarding bridge, and it is very unlikely that a fire would occur, under the precondition and sound fire safety systems like the automatic sprinkler system are installed in other areas, it is recommended that no automatic sprinkler system be installed above the ramps, and the through holes need to have smoke curtains.

The fire detection alarm system, indoor fire hydrant and hose reel system, mobile fire extinguisher, evacuation guidance system, and emergency lighting system on the +8.55 m level should be installed according to relevant standards.

The fire detection alarm system, indoor fire hydrant and hose reel system, mobile fire extinguisher, evacuation guidance system, and emergency lighting system on the +12.15 m level should be installed according to relevant standards.

The fire detection alarm system, automatic sprinkler system, indoor fire hydrant and hose reel system, mobile fire extinguisher, evacuation guidance system, and emergency lighting system on the +17.15, +20.65, and + 21.65 m levels should be installed according to relevant standards.

5.3   Qualitative and Quantitative Analysis

In the design, the approach of fire safety performance-based design and assessment was adopted to analyze areas like people evacuation safety and fire spread at the Western Terminal of Shanghai Hongqiao Transportation Hub and develop fire safety plans for the exit level, platform level, waiting level, waiting hall, and elevated mezzanine. Furthermore, through CFD smoke simulation as well as simulation and analysis of people evacuation in case of fire, fire safety strategies were proposed for the Western Terminal.

5.3.1   Hazard Source Identification and Fire Hazard Analysis

The method of preliminary hazard analysis is used for hazard source identification and fire hazard analysis in various areas of the project, as shown in Table 5.1:

5.3.2   Designed Fires

Designing fires are generally divided into fires with heat release rate (HRR) increasing with time and fires with constant HRR. For smoke exhaust volume determination and structural analysis, fires with constant HRR are considered, while for field model analysis of smoke flow, fires with increasing HRR (t^2 fires) are chosen. Designing fires include the determination of fire size and growth rate.

For fires with increasing HRR, as the time constant for the fire growth of different combustible substances is different, according to the speed of HRR increase, the

Table 5.1 Hazard source identification and fire hazard analysis in various areas

Location	Hazard analysis	Remarks
Waiting area in the train station and at the airport	Possible fire in the area includes: (1) Luggage fire: carry-on luggage, low fire load, low fire hazard (2) Seat fire (3) Commodity fire (4) Catering fire	Restrict commercial activities and recommend fire bulkhead or fuel island for fire safety
Entry/exit passenger activity platform	Possible fire in the area includes: (1) Luggage fire: carry-on luggage, low fire load, low fire hazard (2) Seat fire (3) Commodity fire (4) Catering fire	Restrict commercial activities, centralize commercial activities, and set up separate fire compartments

time squared fire is usually divided into four categories, namely, super-fast, fast, medium-speed, and slow fire.

(1) For luggage fire, normal carry-on luggage fire does not exceed 0.5 MW (an experiment of British Building Research Establishment showed that the biggest size of fire caused by two pieces of carry-on luggage burning simultaneously is 500 kW); referencing the performance-based fire safety design of other passenger train stations and considering a certain safety coefficient, the biggest size of multiple pieces of luggage is 1.5 MW.

(2) For kiosk fire, when there is no automatic fire extinguishing system in place, according to TM19, the maximum HRR per unit for the commercial area is 500 kW/m², so the biggest fire size for a 10 m² is 5.0 MW. And for areas with relatively low flammable load density like the information desk and the service desk, according to CIBSE GUIDE E, the HRR per unit of 290 and 249 kW/m² for office and hotel guest room is respectively adopted. It is recommended in this project that the fire size of the kiosks should be controlled at 3.0 MW (fast fire), with the construction area and the spacing no greater than 20 m² and 8 m, respectively.

(3) For shop fire, the design concept of "fire bulkhead" is adopted. For shops with an area not greater than 100 m², according to the analysis result shown on following table as well as relevant literature, a fire size of 2.0 MW (fast fire) is selected when the sprinkler system (fast response nozzle) functions well; in case of sprinkler failure, a fire size of 6.0 MW is selected for smoke flow and control analysis.

(4) For seat fire in the general waiting area, currently metal frames are adopted in the general waiting area, and there is some soft filler on the seat bottom and backrest. According to SFPE Fire Safety Engineering Manual, the maximum HRR of a seat with metal frame and not much soft filler on the bottom and

backrest is 280 kW, and the fire growth speed is 0.0086, somewhere between slow- and medium-speed fire. As the frame is metal, the possibility of fire spreading is low.

In the smoke exhaust design and people evacuation safety assessment, conservatively consider seven ordinary seats burning simultaneously (without sprinkler), so the biggest fire size is 2.0 MW, falling in the category of medium-speed fire.

(5) For seat fire in the VIP/Business waiting area, there is normally furniture including sofas and tea tables in this area. Sofas usually use wood frames, and the cushion and backrest generally are made of textile or leather with combustible foam material. It is recommended that tea tables shall use non-combustible material to control fire spread. Without considering the function of the sprinkler system, the fire size of the seats in the VIP/Business waiting area is far greater than that of ordinary seats. The fire size of public place (without the sprinkler system) as shown in Table A4-4 (8 MW) is conservatively selected as the scale of the seat fire in the VIP/Business waiting area.

(6) For public places without the sprinkler system, such as the open catering areas on the mezzanine, the biggest fire size of 8 MW in Table 5.2 is selected (kitchens are handled as fire safety units) for structure analysis. Based on the aforesaid, the fire categories and scale of the public areas in the project are determined as shown in Table 5.2.

Table 5.2 Fire category and scale determination of public area

No.	Fire category	Fire class	Maximum heat release volume Q(MW)		Remarks
			With valid sprinkler system	Without sprinkler system or sprinkler failure	
1	Kiosk fire	Fast	2.0	3.0	The size of the fuel island should be controlled within 20 m²
2	Shop fire	Medium	2.0	6.0	Fast response nozzle should be adopted for the sprinkler system in the fire bulkhead
3	Luggage fire	Medium	—	1.5	Multiple pieces of luggage are considered
4	General seat fire	Medium	—	2.0	Metal frame (there is small amount of filler in the seat and the backrest)
5	Seat fire in the VIP/business waiting area	Fast	4.0	8.0	Wood frame (there is more filler in the seat and the backrest)
6	Public space	Fast	4.0	8.0	Open catering area uses 8.0 MW for control

5.3.3   Fire Scenario Setup

Based on the descriptions above of some design fire scenarios in the fire safety performance-based design, considering the fire hazard possibility of different functional areas and different combustible substances at the Western Terminal of Shanghai Hongqiao Integrated Transportation Hub, and according to the principle of "credible and most dangerous fire scenarios," the fire scenario design of the Western Terminal is as follows (Table 5.3):

Table 5.3   Fire size of various main areas at the Western Terminal

Elevation (m)	Specific location	Floor height (m)	Ceiling height (m)	Function or combustible substance	Fire growth rate	Spray control fire (MW)	Design fire size (MW)	Assessment method
−7.95	Self-service check-in hall	7.95	4.80	Large luggage	Fast	1.1	1.7	CFAST simulation
±0.00	Luggage claim hall	12.15	8.25	Luggage pile	Fast	1.9	2.9	FDS simulation
		12.15	8.25	—	Steady state	Not considered	8.0	Steel structure analysis
±0.00	Luggage processing hall	12.15	8.25	Luggage pile	Fast	1.9	2.9	CFAST simulation
±0.00	Concourse waiting hall for remote stands	8.55	6.00	Carry-on luggage	Fast	1.4	2.1	Smoke exhaust calculation
±0.00	Concourse arrival hall for remote stands	8.55	6.00	Carry-on luggage	Fast	1.4	2.1	Smoke exhaust calculation
±0.00	Concourse CIP lounge	8.55	6.00	Part of the area within independent fire safety subzone	Fast	1.4	2.1	Smoke exhaust calculation
+4.20	Flight connection area	4.35	3.00	Carry-on luggage	Fast	0.7	1.1	CFAST simulation
		4.35	3.00	Shops	Fast	0.7	1.1	Smoke exhaust calculation
+4.20	Concourse waiting area	4.35	3.00	Carry-on luggage	Fast	0.7	1.1	CFAST simulation
		4.35	3.00	Shops	Fast	0.7	1.1	Smoke exhaust calculation

(continued)

Table 5.3 (continued)

Elevation (m)	Specific location	Floor height (m)	Ceiling height (m)	Function or combustible substance	Fire growth rate	Spray control fire (MW)	Design fire size (MW)	Assessment method
+5.25	Arrival mezzanine	6.90	3.00	Carry-on luggage	Fast	0.7	1.1	Smoke exhaust calculation
+8.55	Concourse waiting area	9.60	7.20/9.60	Carry-on luggage, open commercial area	Fast	1.8	2.7	CFAST simulation
+8.55	Concourse corner area	12.10	9.60/12.10	Carry-on luggage, open commercial area	Fast	3.0	4.5	FDS simulation
		5.00	2.8	Shops, restaurants, VIP area, children's play area	Fast	0.7	1.1	Smoke exhaust calculation
+13.55	CIP, VIP area	8.10	8.10	Tables and chairs	Fast	1.9	2.9	Smoke exhaust calculation
		8.10	8.10	—	Fast	Not considered	4.5	Steel structure analysis
+12.15	Check-in hall	11.65	11.65	Large luggage	Fast	2.9	4.4	FDS simulation
		5.00	3.60	VIP lounge	Fast	0.8	1.2	Smoke exhaust calculation
		11.65	11.65	—	Steady state	Not considered	8.0	Steel structure analysis
+12.15	Security inspection hall	8.50	5.70/8.50	Carry-on luggage	Fast	2.0	3.0	CFAST simulation
+36.65	Roof corridor	—	—	Carry-on luggage	Fast	No sprinkler	1.5	Steel structure analysis

5.3.4   Evacuation Safety Analysis

1. Calculation method for number of people in various areas of the Western Terminal of Hongqiao Hub (Table 5.4):

2. Consolidation of the number of people in various areas of the Western Terminal (Table 5.5)

Table 5.4 Calculation principle for number of people in various areas of the Western Terminal

	Area		Principle for calculating number of people	Time of stay (minutes)
Western terminal	−7.95 m level	Self-service check-in hall (no luggage)	Traffic flow method	15 (check-in personnel)/5 (non-check-in personnel)
	±0.00 m level	Luggage claim hall	Traffic flow method	30
		Office area within the hall	People density method	—
	±0.00 m level	Luggage processing hall	People density method	—
	±0.00 m level concourse	Arrival area for remote stands	Traffic flow method	10
		Arrival area for remote stands	Traffic flow method	30
		VIP/CIP waiting area	Per number of seats	—
		Commercial and office area etc.	People density method	—
	+5.25 m level	Arrival mezzanine	Traffic flow method	10
	+4.20 m level	Flight connection check-in area	Traffic flow method	20
		Arrival area	Traffic flow method	10
	+4.20 m level concourse	Arrival area	Traffic flow method	10
		Waiting area	Traffic flow method	30
		Commercial and office area etc.	People density method	—
	+8.55 m level – main building	Waiting area	Traffic flow method	30
	+8.55 m level – concourse	Waiting area	Traffic flow method	30
	+12.15 m level	Check-in hall	Traffic flow method	30 (people checking in here)/10 (people checking in in other areas)
		Security inspection hall	Traffic flow method	5
		Office area	People density method	—
	+13.55 m level	CIP, VIP	Per number of seats	—
	+17.15, +20.65 m levels	Office area	People density method	—
	+24.65, +28.65, +32.65, +36.65 m levels	Office area	People density method	—

Table 5.5 Consolidation of number of people at the Western Terminal

	Area		Number of people
Western terminal	−7.95 m level		334
	±0.00 m level – luggage claim hall	±0.00 m level – Area C1	3462
	±0.00 m level – luggage processing hall	±0.00 m level – Area C2	471
	±0.00 m level – concourse	±0.00 m level – Area C3, C4	1433
	+5.25 m level arrival mezzanine	+5.25 m level – Area C1	98
	+4.20 m level – flight connection check-in hall	+4.20 m level – Area C2	1293
	+4.20 m level – concourse	+4.20 m level – Area C3	1655
		+4.20 m level – Area C4	1655
	+8.55 m level – waiting hall	+8.55 m level – Area C1	762
		+8.55 m level – Area C2	1440
		+8.55 m level – Area C3	1442
	+12.15 m level – check-in hall	+12.15 m level – Area C1	3168
	+12.15 m level – security inspection hall		788
	+13.55 m level – CIP, VIP	+13.55 m level – Area C2, C3	330
	+17.15 m level		191
	+20.65 m level	+20.65 m level – office	145
		+21.65 m level – restaurant	1795
	+24.65 m level		219
	+28.65 m level		219
	+32.65 m level		193
	+36.65 m level		496

3. The table below shows the time of phased evacuation at the Western Terminal of Hongqiao Integrated Transportation Hub, including the detection time, alarm time, people response time, and evacuation time. Considering that in reality there might be some uncertainties that may affect evacuation, such as the evacuation of the disabled and children, a 50% safety coefficient is added to the evacuation time obtained from the simulation (Table 5.6).

4. Fire evacuation safety assessment (Table 5.7)

Table 5.6 Evacuation time summary of the Western Terminal

Area	Detection time (s)	Alarm Time (s)	Pre-evacuation time (s)	Evacuation time (s)		Total evacuation time (s)
				Simulation result	×1.5 safety coefficient	
−7.95 m level	150	10	120	122	183	463
±0.00 m level area C1	200	10	120	157	236	566
±0.00 m level area C2	200	10	60	155	233	503
±0.00 m level area C3/C4	168	10	60	59	89	327
+5.25 m level area C1	114	10	120	70	105	349
+4.20 m level area C2	118	10	120	203	305	553
+4.20 m level area C3/C4	118	10	120	135	203	451
+8.55 m level area C1	192	10	120	390	585	907
+8.55 m level area C2/C3	254	10	120	182	273	657
+12.15 m level area C1	248	10	120	225	338	716
+13.55 m level area C2/C3	198	10	120	188	282	610
+17.15 m level	120	10	60	67	101	291
+20.65 m level – office	116	10	60	120	180	366
+21.65 m level – roof restaurant	136	10	120	131	197	463

Table 5.7 Evacuation safety assessment

Fire scenario	ASET(s)	RSER(s)	Meet safety requirements or not
1–1	>1200s	516	√
1–2	>1200s	516	√
1–3	Entry level >1200s	516	√
	Transfer mezzanine >750 s	461	√
1–4	Entry level >1200s	516	√
	Transfer mezzanine >726 s	461	√
1–5	>1200s	516	√
1–6	>1200s	516	√
2–1	>1200s	384	√
2–2	>1200s	384	√
2–3	>1200s	338	√
2–4	>1200s	338	√

5.4   Summary of the Chapter

This chapter takes the terminal of Shanghai Hongqiao Integrated Transportation Hub as an example to introduce in detail the application of performance-based design methodology in reality. Firstly, the overview of the project of Hongqiao Transportation Hub is presented, and the fire safety difficulties caused by its architectural characteristics are summarized, relating to which the performance-based fire safety design is introduced. Following that, tailored fire safety strategies are developed in the areas of fire compartmentation, fire separation, smoke control, and safe evacuation, among which, for smoke control and safe evacuation, quantitative analysis with software simulation is adopted to verify the effectiveness of the smoke control strategy and evacuation strategy.

Chapter 6
Application of New Technologies in Fire Safety for Large Transportation Hubs

6.1 Application of BIM Technology in Fire Engineering

BIM (Building Information Modeling) technology means creating a reusable 3D digital model that integrates project entities and functions. The information of the building life cycle including design, construction, and operation is integrated on the same platform to realize intuitive expression, collaborative operation, information sharing, and data exchange. With this technology, it is possible to predict various problems that will occur in the later construction and operation phases in the early design stage, integrate and deal with hidden dangers in advance, reduce unnecessary resource consumption, and improve management efficiency and safety in the project life cycle. After successful trials, the technology has received extensive attention and been well applied in the engineering field. Large transportation hubs feature crowded people, high fire safety levels, complete fire control systems, and high degree of coordination among systems, and pipelines should be arranged comprehensively and reasonably without collision. These features cause difficulties for fire safety design, construction, and later operation, maintenance, and management, and these problems can be rightly solved via BIM technology [14].

6.1.1 Application of BIM in Fire Safety Design Stage

1. BIM establishment and detailed design. Fire engineering systems are complex and have many pipelines, putting higher requirements on the design and coordination of various disciplines. Therefore, it is easy to cause drawing contradictions and design errors, which can seriously affect design efficiency and quality. BIM technology, however, can be used to enter drawings of various disciplines into the BIM system. Then, based on the actual situation on the site, it is possible to determine the optimal design plan and establish a 3D model of the project.

© Springer Nature Switzerland AG 2021
F. Li, H. Li, *Fire Protection Engineering Applications for Large Transportation Systems in China*, https://doi.org/10.1007/978-3-030-58369-9_6

Due to the intuitiveness of the 3D model, it is possible to realize smooth coordination among various disciplines, clear pipeline design, reasonable and orderly layout, accurate fixing positions of supports and pendants, and reserved maintenance space, which can effectively reduce design errors, greatly improve construction quality and efficiency, reduce construction risks, and facilitate later maintenance, laying a solid foundation for completing the project as planned and controlling the investment. The application of BIM in the structural deformation of Beijing Daxing International Airport is as follows (Fig. 6.1):

2. Collision testing and simulation technology disclosure. (1) Fire engineering pipelines are intricate. The decorative drawings are designed by different units, so there may be problems such as collisions and design errors, resulting in rework and economic losses. After the 3D model is established, the system can perform collision test to generate collision report and reserved opening report, so that we can accurately and intuitively understand the position of each collision component and then optimize the design. After various disciplines adjust the elevation and orientation of the corresponding components for several times, the collision points can be gradually reduced and finally disappear. (2) BIM software can output the optimized 2D construction drawing of the decoration discipline according to the final 3D model, guiding the collaborative construction of decoration and fire-fighting projects. (3) Technology disclosure through digital visualization model virtual roaming technology is also possible. By doing so, the participating parties can intuitively and comprehensively understand the internal conditions of the project before starting construction, thus effectively improving communication and work efficiency and reducing differences in individual understanding of drawings. Especially in terms of disclosure of node technolo-

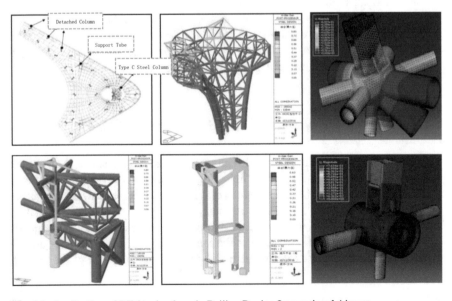

Fig. 6.1 Application of BIM technology in Beijing Daxing International Airport

gies featured by cross-discipline and complex directions, it is possible to jointly overcome important and difficult points in construction and improve the quality of construction, with remarkable effects.

3. Application of BIM in engineering cost and cost control. BIM combines five-dimensional parameters related to cost, such as time, space, and process. In the process of establishing a 3D model, detailed project characteristic parameters are entered for all components, such as specifications, models, materials, size, and positions. After the modeling is completed, the model can be directly imported into the calculation software on the same platform to quickly generate accurate and detailed bill of quantities. This can greatly save the workload and accelerate the progress of the project.

6.1.2   Application of BIM in Construction Stage

1. Management of flow construction organization. Fire engineering is a part of new construction, so it is necessary to organize the construction with professional work types. Besides, the existing construction sites feature multiple restrictions and insufficient space, so construction arrangement and flow organization are difficult. Traditional construction organizations make judgments by virtue of quotas and experience. The construction conditions cannot be clearly estimated, so the construction plan inevitably has loopholes, and there are no effective measures against emergencies. These uncertain factors will all affect the smooth development of the project. The BIM with time attribute dimension can realize dynamic simulation construction and effectively control the construction progress, quality, and safety of the project.

2. Management of manpower, materials, and equipment. It is difficult for construction enterprises to manually calculate the planned consumption under multiple conditions such as different regions and schedules in a short period of time. Most of the manpower and material preparation plans can only be based on estimation, and the system of material requisition based on quotas often becomes formalistic. The BIM can split the amount of work according to the schedule, construction sections, professional teams, and other conditions and generate a list of consumption of manpower, materials, and machines. It is also attached with time attributes. By sorting the material lists in time sequence, you can obtain the material procurement approach plan and reasonably draw up the purchasing and warehousing plan.

3. Management of on-site quality and safety. The BIM technology can use mobile terminals to collect site photos and upload them at once. It can be seamlessly linked to the BIM to timely discover brutal construction, quality defects, security risks, concealed work archive storage, etc., realizing visualization of defects and concealed works. What is more, it can facilitate the supervision of the rectification of construction problems.

6.1.3   Application of BIM in Acceptance and Cost Accounting Stage

Before completion settlement, the data in the BIM is used as the basis for the settlement strategy to reduce missing items and omissions and increase the settlement cost within a reasonable range. During supplier settlement, the model is used to clearly define the settlement scope of each supplier, reduce "loss of profits" in the settlement process, easily deal with settlements, and improve the efficiency of completion cost control (Fig. 6.2).

6.1.4   Application of BIM in Operation and Maintenance Stage

The particularity of fire engineering determines the extreme importance of its operation and maintenance. Personal safety is the first priority at any time. Preventing safety accidents from happening or carrying out emergency response to solve the problem after an accident is the standard for testing the operation and maintenance of fire engineering. The 3D BIM with fully integrated construction information data is the magic weapon for the owner's later operation and maintenance, as well as a powerful tool for the fire company to undertake subsequent maintenance project.

1. Efficient fire equipment maintenance. Detailed information such as specifications, models, manufacturers, warranty services, maintenance knowledge, etc. of the fire engineering equipment can be imported into the integrated fire control system from the BIM before the equipment is put into use, which is convenient

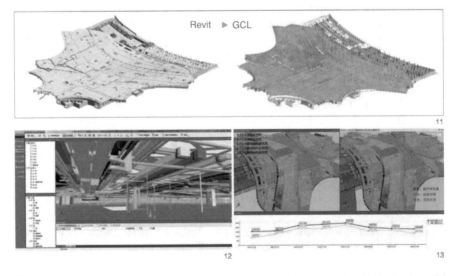

Fig. 6.2 Beijing Daxing International Airport conducts calculation analysis with the Revit model

for later maintenance. Maintenance personnel can analyze and track the running status of the equipment on the platform in time and carry out regular inspection and maintenance to eliminate safety hazards from the very beginning.

2. Rapid fire emergency evacuation simulation. When a fire occurs, BIM can also perform emergency evacuation simulation, quickly calculate the evacuation time according to the data model, select the optimal escape route, and cooperate with the fire department to effectively organize evacuation and reduce the casualty rate.

3. Reduced operating costs of fire safety system. The BIM system can comprehensively collect the operating energy consumption data of fire safety equipment, conduct sorting and analysis, and find ways to save energy and reduce consumption. Through system intelligent monitoring, it can also find equipment abnormalities in real time and save operating costs.

6.2 Application of IoT Technology in Fire Safety

The Internet of Things (IoT) is a network of objects that connects any object to the Internet through various information sensing devices such as radio frequency identification devices (RFID), infrared sensors, global positioning systems, laser scanning devices, etc. according to a contractual agreement, with the goal of information exchange and communication to realize intelligent identification, location, tracking, monitoring, and management. Its essence is to realize "dialogue" between people and objects and "communication" between objects via the Internet and real-time monitoring and dynamic management of the object environment [15].

6.2.1 Application of IoT Technology in Fire Safety Facility Management

At present, fire hydrants, automatic sprinklers, fire water cannons, gas fire extinguishers, automatic fire alarm facilities, etc. installed in large transportation hubs have been connected through fire alarm linkage controllers to form the IoT of independent fire safety facilities of each building. However, the information collection is still not perfect, and monitoring chips need to be implanted in the fire safety equipment to collect and transmit information such as fire safety power supply and pipeline pressure and realize the remote control of the fire safety linkage equipment. If sensing devices are installed on fire hydrants, sprinkler system pipe network, pump group, etc., the pipeline water supply pressure and the working state of the pump group can be obtained in time. When the pipeline pressure drops, it can be detected timely, and fire pump group boosting can be remotely activated via the control chip, so as to ensure that the water supply pressure of the fire pipe network is always up to standard. On this basis, it is possible to connect independent fire safety facilities of each building to a global fire safety facility IoT via the Internet, create a centralized control center system and a central database, and build an operating platform which carries out centralized control, multi-point supervision, and remote control of fire safety facilities, with the goal of realizing full-time and comprehensive supervision of fire safety facilities, effectively preventing equipment failures, timely eliminating fire hazards, and ensuring that the fire safety facilities of the transportation hub are always in good condition (Fig. 6.3).

6.2.2 Application of IoT Technology in Fire Safety Management

Firstly, fire inspection. QR code electronic tags are installed in fire safety facilities and key places of the transportation hub. During inspection, the staff uses mobile phone scanning to synchronously transfer the information to system database, managers' mobile phones, and other management terminals through the network. This can guide the staff to master checking points, checking contents, and testing methods of facilities and equipment, realize fire inspection of equipment and places, and enable the management to keep abreast of the implementation of the duties of

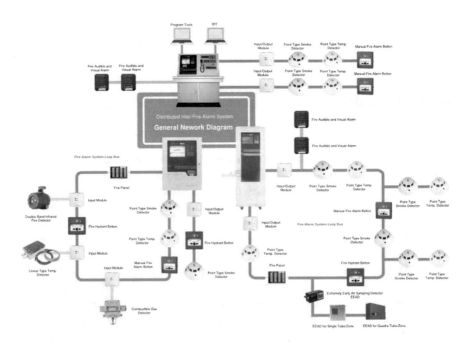

Fig. 6.3 Application of IoT in fire safety facility management

inspectors. Meanwhile, an inspection account is automatically generated in the background terminal, which improves the work efficiency.

Secondly, supervision of key locations. The remote visual image transmission system is built by using the IoT, and image monitoring devices are installed in key locations such as fire control room and fire pump room to monitor the personnel in the fire control room in real time and keep abreast of the fire safety status of key places such as the fire pump room. With the help of the IoT image transmission system, it is possible to solve problems that are difficult to find in the daily fire inspection of the "four-electricity" rooms along high-speed railways and implement daily remote video fire inspection.

Thirdly, daily fire management. The fire safety IoT centralized control center system is installed in the hub unit and the fire system maintenance unit, and an operation platform featuring centralized control, multi-point supervision, and remote control is built to timely transfer the information of fire safety facilities and fire management of the transportation hub to the computers of fire management departments at all levels and the mobile phones of managers. Through mobile phones, the management can check the fire prevention work of high-speed railways, master the status of fire safety facilities, promptly solve the problems found in the inspection, and eliminate on-site fire alarms and fire safety facility failures (Fig. 6.4).

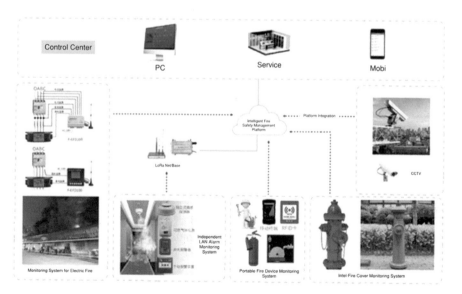

Fig. 6.4 Application of IoT in fire safety management

6.2.3 Application of IoT Technology in Fire Supervision

Firstly, mastery of information about the units under jurisdiction. The fire department sets up a fire safety IoT supervision platform for the transportation hub and transmits in real time the information of the fire safety facility status, fire management status, and fire system maintenance status to the fire department, so that the fire department can fully grasp the fire safety situation of the transportation hub.

Secondly, daily fire supervision and check. Using the remote monitoring function of the IoT, it is possible to conduct real-time spot checks on fire safety facilities and equipment, employees' fulfillment of fire safety duties, on-site fire management, etc.; effectively perform fire supervision duties; save police power, vehicles, and other consumption; and focus the fire supervision on key places with large fire risks and more fire hazards.

Thirdly, investigation and punishment of fire safety violations. For fire hazards and fire safety violations, the fire department uses the IoT operation platform shared by the hub unit, the maintenance unit, and the fire department to establish a hazard supervision operation platform which can directly issue rectification notices to responsible units, clarify rectification responsibilities and measures, and improve the efficiency of fire supervision and law enforcement.

Fourthly, fire safety risk assessment. The fire department uses IoT functions of statistics, analysis, and evaluation to regularly analyze the fire safety status of the hub, assess the fire safety facility maintenance and management capabilities of the fire system maintenance unit, evaluate the fire safety status of the hub unit, find the main hazards and fire risks of the fire safety of the transportation hub, clarify the key points of the fire safety work, and provide a scientific basis for strengthening and improving the fire safety work of the hub.

6.2.4 Application of IoT Technology in Fire Emergency Response

Once a fire occurs in a traffic hub, it will bring great losses to the people's lives and property and have a serious impact on society. As a result, it is necessary to rely on professionals with skills and disposal experience to effectively prevent and deal with all types of fire hazards.

First of all, fire technology based on Internet of Things, through configuring a large number of fire alarm, video surveillance, multi-functional sensing, and other equipment in the fire scene, can achieve accurate detection on temperature, wind speed, smoke concentration, toxic and harmful gas concentration, and other information on the site where fire takes place and timely send information to the department responsible for commanding and dispatching and fire department as well as skilled personnel to provide comprehensive and accurate fire field information.

Secondly, combined with GIS technology and indoor positioning technology, it can play an important role in the formulation of fire-fighting and rescue plans.

GIS technology can provide geographical location imaging and even stereo-scopic imaging. In the formulation of actual fire prevention plan, fire-fighting command personnel need to analyze the situation of the entire fire point from the air, and combined with the situation around the fire site, as well as the location water source, road distribution, vehicle dispatch, surrounding buildings, crowd distribution, and other factors to develop a set of feasible plan to fight against the fire and to withdraw (Fig. 6.5).

Indoor, based on indoor positioning technology and building information modeling, the location of personnel can be shown on the building plan so as to assist fire-fighting and rescue personnel to find trapped people. Rescue personnel can also view the location of personnel and find an evacuation and rescue routes through the handheld terminal APP in real time (Fig. 6.6).

Secondly, with the help of the Internet of Things port, command personnel can grasp information on the condition of the fire, fire facilities, fire rescue, and others no matter where they are. They can also use the Internet of Things control platform to directly start fire pump group, smoke-venting machine, and other fire-fighting facilities; issue evacuation, fire rescue, and other instructions; and command the on-site staff to carry out fire emergency disposal and ensure that fire rescue operations be conducted in a timely and efficient manner (Fig. 6.7).

In addition, after the end of fire rescue, the fire department can also rely on the Internet of Things to collect fire information, accurately find the fire point and objects leading to the fire, and provide a scientific basis for finding the cause of the fire and analysis of accident liability.

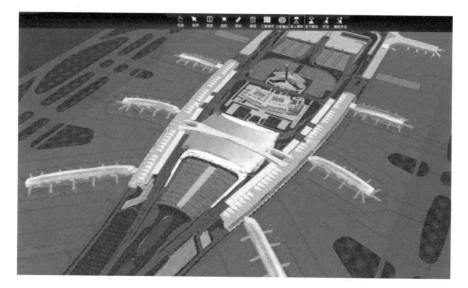

Fig. 6.5 3D imaging of GIS geographic information at an airport

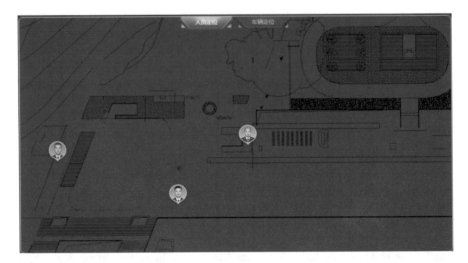

Fig. 6.6 A software for indoor positioning technology

Fig. 6.7 Rescue and evacuation platform on a software

6.3 Safety Protection and Fire Safety in Transportation Hubs

The safety management of large transportation hubs mainly includes fire safety management and safety protection management. Fire safety management aims to prevent fires, reduce losses and casualties, and resist damage from fires. Safety protection management is dominated by protection against theft, damage, explosion, and other social humanistic damage. There are often conflicts and contradictions between these two types of management, but their essence is the same, i.e., to ensure safety. With the rapid development of economic and social technologies, more and more large comprehensive transportation hubs have emerged. These buildings often have multiple business types, and their management requirements and standards are difficult to unify, which brings great difficulties to building safety management.

1. Contradictions between security and fire safety

Large transportation hubs are public places where people gather. Once a fire breaks out, it may cause mass casualties and serious social impact. The investigation results of fire accidents show that although fire accidents with high casualties are due to multiple factors such as imperfect fire safety system, failure to implement fire safety responsibility system, and incomplete fire safety facilities, over 80% of these accidents are directly caused by reasons such as the safety exit doors are locked, the evacuation passage is blocked, and the smoke exhaust windows are closed, so when the fire occurs, people cannot escape, and the smoke cannot be discharged. In many rescues, the fire-fighters have to first remove the security doors and iron fence windows before they can rescue the trapped people, thus missing the best time to extinguish the fire. The contradiction between security management and fire safety management has been difficult to be effectively resolved and has become a stubborn problem throughout the country. For example, security management requires that windows should be less opened and kept closed, especially for shopping malls and museums where there should be no windows. On the other hand, fire safety management requires that doors should be opened, as well as windows. This creates a contradiction. Combining security management with fire safety management through technical means to overcome the long-term contradiction is of great significance for saving people's lives and protecting property safety.

2. Discussions on the integration of daily management of security and fire safety

(1) The management center of the fire automatic alarm system and the safety protection monitoring system is the central control room. For the convenience of management, it is recommended to build the security control room and the fire safety control room together, thus forming a structure of more than two systems with compound control capabilities, i.e., a comprehensive control room. The following recommendations are on staff structure and management:

Centralized management of comprehensive control room personnel:

After the security control room and the fire safety control room of a building are integrated into one monitoring room, it has both security and fire safety functions, but it is still recommended to arrange a group of personnel to be on duty. The management must be dominated by fire safety. It is required to use security surveillance cameras to quickly grasp the situation of the alarm site and identify whether there is a fire, with the goal of shortening the fire handling time and realizing the effective use of human resources while making full use of the security and fire safety equipment.

Improving the competence of the management personnel in the comprehensive control room:

No matter how advanced the equipment in the comprehensive control room is, people are needed to operate it, so it is fundamental to establish a stable and reliable team of comprehensive control room personnel on duty. The comprehensive control room should be equipped with monitoring personnel of a high education level, and the supervisors are required to master the operating procedures of all equipment in the comprehensive control room. In addition, social units should regularly hold short-term training on laws and regulations for fire control room personnel on duty. The personnel should study *Fire Safety Law of the People's Republic of China*, *Fire Control Room Management and Emergency Procedures*, *Maintenance Management for Building Fire Equipment*, *General Technical Requirements for Fire Control Center*, and other related laws and regulations, technical specifications, and fire cases and increase the sense of responsibility and honor of the position from the heart.

(2) Construction of the technology platform

New automatic alarm system integrating fire safety, security, and IoT:

Nowadays, computer technology, communication technology, and electronic information technology provide reliable technical support for the effective combination of security surveillance systems, access control systems, and video intercom systems in security systems and safety protection systems. If security, safety protection, and IoT are actively and effectively integrated in a building and applied in the automatic fire alarm system, then the new automatic alarm with prevention as the priority and with the combination of prevention and elimination as the guideline will play a huge role. When a fire occurs in a place and an alarm is given, the automatic fire alarm system is linked to the video surveillance system to call out the images of relevant video monitoring points at the fire scene, enabling the personnel to see the fire situation and the flow of people in real time, and judge the authenticity of the fire alarm. The linkage between the access control system and the video intercom system plays a key role in the evacuation of people and the positioning of trapped people. After a fire alarm is given, the access control system opens all the locked doors, so that the safety exits are open to guarantee the evacuation of people. The most important thing is that when an accident occurs, the important information is immediately provided to the rescuers, thus effectively reducing the rescue time. This new type of automatic alarm system integrating security, fire safety, and IoT is a prerequisite for the safety management of social units, and it will also be a useful tool for obtaining reliable data during accident rescue and accident handling.

Applications of new products:

Fire image recognition system: When using a traditional image monitoring system, the security personnel have to observe the images to determine whether there is any abnormality (such as fire and theft) in the monitored place, which means a lot of work for the security staff. Even so, it is difficult to ensure uninterrupted surveillance 24 h a day. With the development of science and technology, computer hardware and software have been able to meet the requirements of developing camera-embedded applications. Fire image recognition is to use cameras covering all areas of the building and a set of image recognition software installed in the video surveillance system to automatically detect fire characteristics such as smoke, fire, and white gas in the video scene. The system can automatically detect a fire breaking out in the monitored area and give an alarm. The entire video surveillance system uses computers to replace the observation of images by people, so it can greatly reduce the labor intensity of the security personnel. What is more, as an auxiliary monitoring system of the automatic fire alarm system, the system has a much more detection time than the automatic fire alarm system, so it can realize early fire detection and avoid secondary disasters such as misoperation of the spray system.

It is understood that relevant video surveillance manufacturers have been researching the embedding of fire detection video algorithms into a camera that can detect fires and provide automatic insurance and safety notice. The camera can detect a 75×75 cm fire 840 m away in 5 s. As for a 6×6 m fire, the detection distance can be further. In addition, the camera sets the temperature threshold. When scanning the monitored area, the camera will give an alarm if it detects that any object in the area has reached the critical temperature.

Real-time monitoring of people: For densely populated places in transportation hubs, real-time monitoring of the number of people is of great significance for the formulation of safety management plans for these sites. Traditional fire management often ignores the monitoring of the number of people, mainly due to the lack of practical people monitoring solutions. At present, the safety protection system can detect passenger flow through both video surveillance and infrared radiation. The video passenger flow detection system tracks moving objects in a single or multiple video surveillance areas based on intelligent algorithms such as artificial neural network and pipe diameter feature matching and by using big data and cloud computing. It detects and recognizes a moving object according to the principle and method of pattern recognition to determine whether it is a human body, thus accurately detecting the number of customers passing through the area. As for the infrared radiation passenger flow detection system, infrared radiation detectors are installed at the main entrances and exits of the building to calculate the number of people entering and exiting, thus calculating the number of people inside the building.

If the automatic fire alarm system can be combined with the safety protection passenger flow detection system, the automatic fire alarm system can actively monitor the number of people in densely populated places in real time. When the number of people reaches the maximum allowable value for safety exits, an alarm should be issued to remind the management personnel to take measures to control the number of people. By monitoring the number of people, it is also possible to draw a distribution map of people inside the building, which helps social units to formulate corresponding fire emergency plans and drill plans accordingly.

Application of BIM: The core element of intelligent building construction lies in integration. In recent years, BIM technology has also developed rapidly. By setting the integration of security system and fire safety system on the building information modeling, it is possible to not only realize a visualized intelligent building model and break through the traditional security and fire safety management based on 2D drawings but also build an intelligent management platform to improve management efficiency. Through the visual management platform, it is possible to present the building's spatial information, fire safety equipment, safety protection equipment, etc. in a 3D top view and integrate fire safety, security, building model, and facilities and equipment into a large platform for management. As a result, we can better coordinate the interlocking actions and mutual cooperation among subsystems to achieve unified management, greatly simplify the occupation of system resources, and ultimately improve the efficiency of management.

6.4   UAV Fire Extinguishing Technology

As a new type of industrial technology, the UAV fire safety technology has been widely used in various fields. In China, many fire safety organizations have successfully conducted fire site detection and monitoring and airdrop of relief materials using UAVs, with very obvious effects. In the rescue following the Tianjin Port explosion accident, various departments also used UAVs to conduct high-altitude detection on the scene of the accident, providing some reference for rescue decision-making.

With the overall solution using accurate, intuitive, and comprehensive fire site analysis data, rescuers on the scene can obtain timely information. There are few multi-function fire safety UAVs on this basis, especially fire safety UAVs which can still perform tasks near the fire site. Because of technical, engineering, and financial problems, few companies are engaged in them. Experts believe that with the popularization of technology, UAVs have gradually gained attention and industrial-grade UAVs have also gradually been applied to various fields and played their own role. Especially in high-risk areas dominated by fire safety, using UAVs to replace some manual operations is very important. Therefore, we specifically interviewed some companies specializing in industrial-grade UAVs. We learned the importance and application of UAVs in the field of fire safety and that UAVs can play many roles in fire-fighting force considering its actual needs. Currently NFPA 2400 Standard is used for Small Unmanned Aircraft Systems (sUAS) is used for Public Safety (Fig. 6.8).

Fig. 6.8 UAVs used in fire extinguishment

6.5   Chapter Summary

This chapter mainly introduces the new technologies applied in the fire engineering of large transportation hubs: BIM, fire safety IoT, combination of security management with fire safety, and new fire safety technologies. Regarding the use of BIM technology, this chapter expounds the application of BIM technology in the fire engineering of large transportation hubs in four stages, i.e., design, construction, acceptance, and later operation and maintenance, and demonstrates the superiority of multi-discipline collaborative operation, information sharing, and data exchange embodied by BIM technology in the whole life cycle of the building, with Beijing Daxing International Airport terminal as an example. With regard to the application of fire safety IoT technology, this chapter discusses its application in the later stage of construction from four aspects, i.e., fire safety facility management, fire safety management, fire supervision, and fire emergency response.

Chapter 7
Conclusions and Prospects

The first part of this book summarizes the development of the world's large transportation hubs and the evolution of their architectural forms and concludes the transformation of the function of transportation hubs to transfer center of a city's internal and external traffic, the development trend from waiting to passing, the development trend of building space from plane functional space to three-dimensional functional space, the transformation from single-function space to compound space, and the shift from indirect transfer to zero-distance transfer. Due to the superposition of traffic and commercial service functions, this compound space has the characteristics of large plane area and high space. The difficult point in fire prevention design is that there are over-specification situations in fire compartment, smoke control and evacuation, etc., which is also the focus of this paper.

The second part firstly introduces the transfer organization methods and streamlines organization inside large transportation hubs, as well as functional space of large transportation hubs. Then, from the fire safety perspective, it extracts typical space types of large transportation hubs and expounds the fire safety characteristics of large transportation hubs in terms of characteristics of architectural space, fire hazards of building components, and human factor.

The third part, starting from the development history of fire prevention codes for architectural design of large transportation hubs in China and the United States, compares the similarities and differences between the codes of the two countries and provides a code basis for subsequent fire safety design. It then introduces performance-based fire prevention design process and performance-based fire prevention design goals of large transportation hubs and summarizes performance-based fire prevention design methods.

The fourth part, starting from the characteristics of large transportation hubs, extracts the difficulties of fire prevention design in large transportation hubs and actively introduces some important concepts of performance-based fire prevention design for these difficulties. Combined with performance-based design methods, it analyzes and studies the fire prevention characteristics of typical fire prevention

© Springer Nature Switzerland AG 2021
F. Li, H. Li, *Fire Protection Engineering Applications for Large Transportation Systems in China*, https://doi.org/10.1007/978-3-030-58369-9_7

spaces from four aspects, i.e., building fire prevention, safe evacuation, fire rescue, and fire safety management, and proposes fire safety strategies.

The fifth part summarizes the application of performance-based fire prevention strategies in specific engineering practices through the case study of Shanghai Hongqiao Transportation Hub.

The sixth part explores new technologies applied in the fire engineering of large transportation hubs, with BIM and fire safety IoT as examples, and expounds the practical application of BIM and fire safety IoT from three aspects, i.e., design, construction and acceptance, and operation and maintenance.

Through the discussions in this paper, the following experiences worth learning and promoting in future large transportation hub projects are summarized:

1. Performance-based fire safety design should be introduced at the beginning of the design stage, so that the design schemes and fire safety schemes of various disciplines can be organically combined. The fire safety schemes involve multiple disciplines, and the early introduction of performance-based design can avoid complicated scheme modification in the later stage. Take the evacuation design, for example. The evacuation scheme has a great influence on the plane design of the building. If the evacuation scheme can be determined at an early stage, the plane layout will not be modified significantly due to insufficient width or evacuation problem.

2. The continuous exploration and development of fire engineering science, combustion dynamics, and other related disciplines have a strong guiding significance for solving the complex problem of fire mode and smoke flow process in the high compound space of rail transit hubs.

3. The systematic project of performance-based fire prevention design needs to draw on the design theories and methods at the forefront of various disciplines and summarize current lessons, thus formulating generally applicable performance-based design codes.

4. For designs that have a significant impact on investment, advanced performance-based design methods should be actively introduced to balance economic benefits and safety inputs. For large buildings with steel structure, the steel structure fire safety design usually accounts for a large proportion in the initial investment in fire safety design. Prescription-type fire prevention codes don't take into account the characteristics of buildings, so they may be either not safe or not economical. By contrast, performance-based design methods are tailor-made for buildings. They fully consider possible fire scenarios in the buildings and quantitatively calculate the impact on the structure, thus making the design schemes more scientific and reasonable.

5. For high compound spaces, spaces with vertical through atrium, underground spaces, and dense spaces in large transportation hubs, it is required to adopt targeted fire prevention strategies according to their different spatial characteristics and apply performance-based means to make the fire prevention design more flexible. Fire prevention codes should not restrict the development of functional integration and spatial diversification of large transportation hubs.

With the deepening of the construction of large transportation hubs, the construction of new large transportation hubs has also ushered in a new opportunity. Based on the characteristics of large transportation hubs, this paper analyzes the necessity of adopting performance-based fire safety design in transportation hubs, proposes some design ideas and design methods, and introduces their application by taking Shanghai Hongqiao Transportation Hub as an example. On this basis, further research on the fire prevention performance of large transportation hubs should be carried out to make technical preparations for the fire safety design of such buildings in the future.

References

1. Design of transportation port and hub, Huyongju;
2. Performance based fire protection design evaluation report of Foshan West Railway Station and related off-line station building and platform canopy project of new Guangzhou railway hub, Jianyan Fire Design Performance Evaluation Center Co., Ltd;
3. Integrated development of underground railway station and urban core area, Huyingdong;
4. Study on the design of the interior commercial space of the modern terminal – Taking the terminal T3a of Xi'an Xianyang International Airport as an example, Tiandanmeng
5. Fire protection design and audit details of high-rise civil buildings 100, Guoshulin
6. Research on typical space fire protection design of Railway Comprehensive Transportation Hub Based on performance-based method, Jiahuichao
7. Study on the distribution law of passenger flow period of Wuhan Guangzhou high speed railway, Bitianru
8. Fire prevention and control of large commercial complex, Huangminjie
9. NFPA 130, Standard for fixed guideway transit and passenger rail systems
10. Smoke control and exhaust engineering design of large space building, Zhanghu
11. Performance analysis of fire protection design for large subway station, Zengguobao
12. Feasibility analysis of smoke control mode in subway tunnel, Zhengjili
13. Fire protection design of Shanghai Hongqiao comprehensive traffic hub building, Zhaohualiang
14. Application of BIM Technology in fire engineering, Wangdanjing
15. Application of Internet of things technology in fire safety management of high-speed railway, Qiwei

© Springer Nature Switzerland AG 2021
F. Li, H. Li, *Fire Protection Engineering Applications for Large Transportation
Systems in China*, https://doi.org/10.1007/978-3-030-58369-9

Index

© Springer Nature Switzerland AG 2021
F. Li, H. Li, *Fire Protection Engineering Applications for Large Transportation
Systems in China*, https://doi.org/10.1007/978-3-030-58369-9

Printed in the United States
by Baker & Taylor Publisher Services